The Metaphysics of Protection

a dancing star foundation book

BY MICHAEL CHARLES TOBIAS
AND
JANE GRAY MORRISON

Cover photohgraph: The Victoria crowned pigeon (*Goura victoria*) by Micahel Charles Tobias.
Back cover photograph: Orangutan (*Pongo pygmaeus*) in Kalimantan, Indonesia, by Jane Gray Morrison.

ISBN: 978-1-939116-97-0 ebook
ISBN: 978-1-939116-98-7 print edition

TABLE OF CONTENTS

PUBLISHER'S NOTE

I first met Michael Tobias in 1980 when his manuscript for the book that became Deep Ecology was presented to me in my capacity as senior editor for Harcourt Brace Jovanovich Publishing for possible reprint consideration. I was fascinated by the manuscript and it was the first time I had encountered the concept of Deep Ecology. I left Harcourt in 1981 to found my literary agency Waterside Productions Inc. and did not have the opportunity to publish any of Michael's subsequent books but followed his amazing career as a novelist, ecologist and film maker. When he married Jane Gray Morrision the partnership was set not just for a marriage but for a journey that would bring the world some of the most important ideas related not just to ecology but human survival.

In **The Metaphysics of Protection** the deep ecology of the book is the quest for meaning and purpose as human beings. Michael and Jane discuss such a variety of topics and literary and scholarly visionaries that most readers - myself included - will be challenged to recognize the full range of references and ideas. That in my opinion is one of the great attractions and strengths of the book you are about to read. In this age where everyone is looking for easy, quick solutions, Jane and Michael provide us a challenging treatise that will make us think and perhaps do a little research to be sure we truly comprehend the essence of these essays. All the better. This is a true intellectual odyssey and like all such journeys the rewards are in the struggle to overcome challenges. Everyone who reads **The Metaphysics of Protection** will be wiser and more human for the experience. This is a book with a universal message that appreciates the depths of human thought and striving and provides provocative insights along the way. We will all be better human beings and better stewards of our planet just from the experience of reading this book.

My goal as publisher is to see this book serve as a catalyst to conversation. These topics are important topics. They merit thoughtful consideration. I am proud to call Michael and Jane my friends and honored to publish **The Metaphysics of Protection.**

Bill Gladstone, Publisher

FOREWORD

We are at a critical juncture in our history. We have been there before, but not as a species, and not as a crucial element in the history of life on the planet. This is a time when it behooves us to take a moment to reflect on where we are, how we got here, and where we would be – and should be – going. Tobias and Morrison have undertaken such a reflection, and it is a precious resource for all of us. It prompts us to follow their lead.

A species that is conscious of itself and of its environment may be a great advance in the evolution of life, but it also poses a great risk. It would seem logical that a species that does not, need not, or possibly cannot reflect upon itself and its role in the scheme of things, also will not come to wrong conclusions about who or what it is and what its role is. Such a species – which could theoretically mean every living species on Earth, except the human – exists, it does not ask why it exists and how it should exist. The human species seems compelled and destined to ask these questions, whilst attempting to formulate answers to them. But these answers can be profoundly off the mark. Our species can set off on courses of action – trajectories of its evolution – that endanger not only itself, but its environment. And today the environment of the human species is nothing less than the whole web of life on the planet.

We have set off on a dangerous course because we have not asked ourselves the kind of questions this book asks. And because we have ignored that there are such questions to be asked, and that there can be answers to them. An advance into novelty, as philosopher Alfred North Whitehead termed the process of evolution, when unguided by conscious purpose is a high-risk proposition. This is all the more so for humankind. We act in light of our values and beliefs, and if these are mistaken, or simply obsolete, the risk of disaster hardens into certainty. The world is changing, and it is changing faster than our values and beliefs. Unreflective answers are likely to be outdated answers—dangerously wrong answers.

Yes, as Tobias and Morrison tell us, we need a metaphysics to orient ourselves. It has to be a "meta" physics because it has to penetrate beyond the physical plane in our world. Ours is not only a physical plus biological world, it is also a psychological world. We have brought these elements into the equations that define and determine our life and destiny, and we cannot spirit them away, nor can we afford to ignore them.

Einstein said with good reason that we cannot solve the problems that now face us with the same thinking that gave rise to them. The thinking that gave rise to them is still largely tribal thinking. But since the time tribal thinking first established itself in our collective psyche, the human tribe has grown to planetary dimensions. There is only one tribe now in the world, the tribe of living beings, embedded in the biosphere. We are a tribe embedded and entangled with all other living species and with the integral life-supporting systems of the planet.

Do we need a metaphysics of protection in this tribe? Yes we do: we need protection from ourselves. That is not the standard concept of protection, the concept of defending ourselves from things and people outside—from "the rest of the world." The protection we need is from unreflective action, from failing to ask the key questions and searching for answers to them. Answers that locate ourselves as members of the human tribe which, in turn, is a member of the tribe of all living things.

This book by Tobias and Morrison helps us protect ourselves from our unreflective selves. It must be digested, absorbed, and lived. Moving to this deep level of the metaphysics of protection is needed if the great experiment of the introduction of a conscious species in the symphony of life on the planet is not to end in discord and chaos. It need not do so: humanity is not only a danger to life but also a potential promoter of life.

Is it too soon to think of ourselves as the Conductor of the symphony of life on the planet? Clearly, we must first recognize ourselves as a member of the orchestra. Yet, we wield certain powers and influence the impact of which exceeds that of the other members. We may be a conscious member, able to comprehend the symphony. But do such powers grant us the right, and even impose the obligation, to be the first violinist? The one who ensures that the Conductor's instructions are received by and followed by the rest of the orchestra? At face value, given the mysteries and unimaginable range of Earth's biodiversity, this might strike as sheer arrogance by one, human, species.

There have been several candidates, self-proclaimed Conductors. One who deserves to fill this role is yet to appear. He or she must be one who knows the score, and is faithful to the intentions of the Composer. The Composer is Nature: the creator of the score that holds all life on Earth in the embrace of harmony. There is beauty, intent, and purpose in the score. It needs to be perceived, comprehended, and achieved. We, as allegedly first violinists, can and must play a central role in this complex equation of life. In absence of an obvious Conductor, the orchestra must conduct itself, which it has been doing quite magnificently for several billion years of life's presence on this planet.

Enter humanity, guided along by this proposition that the human species must take on the role of first violinist, given our rapid history and the full impression of our proclivities. How might such musicians ensure that they engage the world with integrity, honesty, as fully aware as possible, so as to bring about the flourishing of life on the planet rather than its demise?

There is still time to seek the meaning and purpose inherent in Nature's symphony. We can be grateful to Tobias and Morrison for calling attention to the threatening cacophony in our world, and the importance of reading and returning to the basic score.

Ervin László

June 2014

Dr. Ervin László is Founder and President of The Club of Budapest, Chancellor of the Giordano Bruno University, Co-founder and Director of the Ervin László Center for Advanced Study, Member of the Hungarian Academy of Science, Fellow of the World Academy of Arts and Sciences, Member of the International Academy of Philosophy of Science, and the author or coauthor of fifty-four books. He obtained his PhD from the Sorbonne, in addition to receiving four honorary PhDs. The recipient of numerous international awards, he has twice been nominated for the Nobel Peace Prize, in 2004 and 2005. He lives in a four hundred year-old former chapel in the hills of Tuscany.

PREFACE

During a long and very challenging and rewarding career in which I worked in different disciplines – ethology, ecology, evolutionary biology, conservation sciences, philosophy of mind and ethics - I've poured over countless works – many seminal - on the importance of protecting our planet from further abuse by humans. But Michael Charles Tobias and Jane Gray Morrison's latest tome, a thought provoking meditation on humanity's current ecological and moral predicament, is by far the most wide-ranging and penetrating treatise I have ever read in this arena.

Following upon their equally bold and broad-ranging works, **Sanctuary: Global Oases of Innocence** (Foreword by Her Majesty the Queen of Bhutan), and **God's Country: The New Zealand Factor** (Introduction by PETA President, Ingrid Newkirk), **The Metaphysics of Protection** is truly groundbreaking: a massive work of pantheistic ecology conveyed in as tight and exhilarating a distillation as one is ever likely to encounter.

As I read through this outstanding book I began thinking of it as a jigsaw puzzle because of its depth and breadth and the strategically uncanny juxtapositions of so myriad a range of underlying human experiences, from those of Hugh Lofting's Dr. Dolittle, the worlds of Peter Rabbit, Winnie the Pooh, and the Wizard of Oz to Pythagoras and Aristotle, Phanerozoic cybernetics 542 million years ago, various masters of the Renaissance, Darwin's circle, JeanPaul Sartre, the French Impressionists and author/philosopher Nikos Kazantzakis, and much more.

All of these pieces, each and every one, are critical to a thesis that is as devastating as it is enlightening. The pieces all fit, and one cannot imagine any one of them being absent from this eloquent line of thinking. How the different pieces would fit together is just as important to Tobias and Morrison as why they must fit together to convey a coherent and pellucid whole; a whole that is masterfully greater than the sum of its many outstanding parts.

We are living on a tired and less resilient planet because there are too many of us and we are takers, rather than givers. As the authors poignantly illuminate there is still time for this and future generations to "re-wild our hearts" and walk the talk of compassion and peaceful coexistence with all beings, human and nonhuman. "Our need to revere and celebrate the Earth," say Tobias and Morrison, "may be crucial to the survival of our species, and that of all others." Indeed, this book drives home a very clear message: we must honor all those other life-forms who share their homes with us because we are an integral part of one magnificent planet and we cannot pretend to live in isolation from the horrific things we have inflicted on the biosphere, and continue to perpetrate.

We are "born to be good" and we have done and continue to do many positive things for this one and only planet. I thoroughly agree with these renowned authors that empathy in practice, and with time, will transcend economic, legal, scientific, and political juggernauts.

The **Metaphysics of Protection** provides critical and persuasive data, eloquent and exhaustively researched arguments, and profound seeds of action, especially for those who have all but grown battle weary and tired (burnt-out) of trying to alter the course of human history's folly; its seemingly intractable opposition to most other life forms.

It is not asking too much to do all we can to allow others to live in peace and safety. Read this seminal book. Read it again. Share it widely. **The Metaphysics of Protection** is a game changer, a life changer.

Dr. Marc Bekoff

Dr. Marc Bekoff, Professor Emeritus of Ecology and Evolutionary Biology, University of Colorado-Boulder; Author and/or Editor of more than 25 works including two editions of **The Encyclopedia of Animal Rights and Animal Welfare**, the four-volume **Encyclopedia of Human-Animal Relationships**, **The Emotional Lives of Animals**, **The Animal Manifesto**, **Rewilding Our Hearts: Building Pathways of Compassion and Coexistence**, **The Ten Trusts** (with Jane Goodall), and **Ignoring Nature No More: The Case For Compassionate Conservation**.

Authors' Introduction

"Whence is this dread of ev'ry creature?
Fly they our figure or our nature?"
(John Gay's **Fables**, Fable XV,
"The Philosopher and the Pheasants,"
London: 1753, Volume 1, p. 58.)

"Compassion Fatigue"

No greater conundrum now faces humanity than its sordid relationship to the vast majority of all other sentient creatures – whales, forests, bees and bats, big cats, small cats, herbivores, benighted smoothtooth blacktipped sharks, and so-called bioinvasives, whose itineraries have largely been meted out by our own inadvertent or deliberate designs; our penchant for synthesizing new chemicals and fouling virtually everything we do, touch and invent; patenting new biological organisms of our own making, and exerting in every quadrant of the biosphere a runaway consumerism, the majority of which stems from vast Holocausts hidden from most of us for our convenience. The combined force of this unnatural colossus of pernicious impacts – from the most remote polar deserts and tropical rain belts, to the concretized intricacies of city human conurbation - adds up and subtracts in ways that collectively have rightly been characterized as the Sixth Extinction Spasm in the whole history of life on Earth, the great tragedy that is the Anthropocene.

But such inordinate power also will of necessity deteriorate as it suspects the missing ingredient in its own hyped-up hubris: humility, that biological clause of ultimate persistence that is inherent to being but one person amid a scintillating world of beings, at large. To be aware of our own arrogance, and carry on as if it were not so, is a mindfulness without precedent. That suspicion, those doubts, widen, turn to illness that is marked by any number of unpleasant symptoms, and eventually wizens and collapses from sheer confusion that cannot, in effect, compensate the loss of meaning, the Self, with any anodyne. The greatest sadness of all is that it would hope to do so and, as this treatise shall purport, would attempt this mending by reversion to those very gentle beings and scenes of much childhood. But it is this lethal fallacy, and remains so, of thinking oneself separate and superior, bereft of the sympathetic between our species and most others, that condemns humanity and assures most human old age of a bitterness. Our violence precludes our ability to make friends with all those mythical creatures everywhere around us who elicit primal reactions of loathing or terror, even unto the last day. If we have learned anything from Nature, it is this audacious and untenable syndrome. Some call it ego. Others, civilization. But it is certainly not wisdom.

Well over seven billion people – poised to become a behemoth of 9.5, 10, as high as 16 billion by the end of this century (demographers have been underestimating the human population explosion every year for decades, indeed, over the course of two centuries) - are zealously pursuing goals that are, in most cases, at some or entire odds with the needs of other creatures and their habitats. Those other life forms across the entire suite of taxons may number as many as 100 million species, each with hundreds, thousands, millions, or billions of individuals. The habitats that are being rapidly desecrated by human wants are their habitats in which to survive, not ours to mow down and ruin without concern of consequence. This jeremiad-like litany has become the great wrestling match of our age. Those black bears and Wisents, giraffes and snow leopards, mountain gorillas and desert tortoises, halibut and sea horses, giant clams and tree frogs, Macaws and what we call weeds, have been there for millions if not tens of millions of years, while we are probably younger as a species than 150,000 years, though our evolutionary beginnings certainly commenced some 50 million years ago. So where does that leave us – ethically, philosophically, in mind and in practice?

This little book seeks to unravel the logic and illogic of our persistent claims to the planet in hopes of settling some accounts that will shed light on both the problems and the possibilities of the human presence here on Earth. Methodical study of this human paradox is not a pretty picture postcard, but the grim and multifaceted profile of an unprecedented illness. The patient, the Earth, is hemorrhaging. Thirty years ago virtually no one spoke of global warming, but rather, came to view aerosol cans as the world's end. Then, after the Montreal Protocol was ratified, ozone depletion started to fade from human memory. Now, the notion of atmospheric CO_2 parts-per-million phraseology is taken for granted. Everybody complains about it, as if it is just some new unpleasant weather trait that started yesterday, with new beach house insurance implications, carbon taxes on gasoline, disappearing Monarch butterflies, but little else. Soon, in spite of annoying flux –a doubling of populations in Nigeria and Algeria, the onus of the future hitting the tropical rain belts as well as snow fall in the Rockies and spruce beetles in Alaska -one can predict habituation to these new catastrophic events, even wildfires, a syndrome that thrives on the economic engines inherent to disaster; that profits in direct relation to the human complacency and ennui that inevitably attends the public's compassion fatigue. There is forgetfulness in so many aftermaths.

Unfortunately, these sociological patterns can never be reconciled with self-destruction, or the massive extinction events occurring, both megafaunal and floral. Awareness, pellucid assessments of data, the will to champion what is right for the Earth, may or may not prevail given natural selection as it articulates a human wrath, but these are our only viable options.

"Beethoven's Ninth"

The five essays presented here encompass numerous historical as well as contemporary individuals, works of art, enigmas and discoveries of science, schools of thought, activist exemplars, quandaries, emergencies and imponderables that, in combination, yield a picture of great promise, as well as cruelties. We are, to be certain, not fashioned from some celestial template that defines human nature, or predetermines it, but from the ongoing result of countless decisions made intuitively, consciously or unconsciously minute by minute. Our choices are important, even if our species may not be. This is a telling note in the biological symphony of which we have, for so long, thought of ourselves as unique and superior to other members. We may not be so foolish as to believe ourselves the Conductor of the symphony, but we have ignored all the other musicians and behaved as if it is our own musical invention.

The 200th anniversary of the first public performance of Beethoven's "Ode to Joy," from the 4th Movement of his 9th and final symphony, will take place in May 2024. 1824 was the same year the French physicist Joseph Fourier first postulated the fundamentals of a greenhouse effect, and the human population hit approximately 1.1 billion. By then, literateur and historian Friedrich Schiller, who had written the hymn "Ode an die Freude" in 1785, from which Beethoven borrowed his text, had long lamented that his poem was "detached from reality" and something of a failure.

These days, 190 years later, more or less, the manifest buzzword being "green," we recognize that greenhouse effect, celebrate Beethoven (and Schiller), but appear to be incapable or unwilling to collectively address the crisis, despite the inspiration that a "9th Symphony" elicits in us. Many years ago the United Nations took a global sample of people and their views on the environment. Nearly 90% of all humans thought of themselves as "environmentalists". Yet, other less ambiguous findings have shown that awareness of issues rarely correlates with our willingness or ability to act upon such awareness.

We are equally aware that the entire biosphere is a symphony, and that the ruination of the natural world by our kind will eventually silence all that remarkable music, of a Beethoven, of the Humpback Whales or the red-eyed Vireo, a bird species that sings some 20,000 songs per day.

Perhaps the extinction this year of some fifteen million discrete populations of other organisms, and unknown thousands of species, will go unnoticed by us, but certainly not by Nature herself. Just as mature canopy forests, mangroves, estuaries, wetlands, tall grass prairies, and countless other ecosystems are chipped away hour by hour, so too, the astounding music of nature gets fainter by the minute, whilst our own chain-saw acoustics work round-the-clock to divest and undermine the great chorales of every ecosystem of their miraculous harmonies.

"Opportunity Costs & Benefits"

We have an amazing opportunity in this generation to cease and desist; to revivify: not only to act fast, wisely, from the heart – which remains largely true to its natural origins and compass, which are good ones – but also to set a steadfast and courageous example in the face of so much needless annihilation. We call this book "The Metaphysics of Protection" because it is, indeed an abstruse and dusty set of conjectures we examine; a philosophical framework of no obvious or demonstrable claim that aims to bring together ideas and aspirations, allegedly quixotic ideals, whose calling has comported with the saving of life, not its destruction; with beauty whose sources are inextricably woven from the magnificence, subtle nuances, ineffable presences of ecosystems that have required over 4 billion years to evolve.

While there are countless other star systems with planets that may also harbor life, it is not our intention to speculate on that which is entirely unknown to human experience. Rather, we want to remind readers of what humans have experienced and thought about, in a sincere effort to convey some enchantment that gives time for reflection and re-evaluation.

Some biologists (ornithologists in particular, such as the remarkable Tim Birkhead) have looked at the startling evolution of suggested or applied anthropomorphism into rigorous scientific theory; from the publication of Niko Tinbergen's groundbreaking **The Study of Instinct** (1951), whose methodological apparatus essentially eschewed the least attribution of human-like emotions to other species, and the far more contemporary evocations inherent to Don Griffin's **The Question of Animal Awareness** (1976), in which "consciousness" loomed large and has since become – in many academic disciplines, no matter how murky the definitions – a veritable given. This is so for birds (forgetting the very unclear ratio of brain size to our as yet unrealized conjectures regarding intelligence, emotion, self-consciousness and empathy in others than ourselves, particularly those precocious social insects among whom bees have recently been shown to have minds of their own, quite apart from the peer pressure of colonies). At the very least, our ethological Weltanschauung takes it as a given the famed lines from the prodigious John Stuart Mill's "The Art of Living" (**Principles of Political Economy with some of their Applications to Social Philosophy**, Part IV,

Chapter VI, 1848): "Nor is there much satisfaction in contemplating the world with nothing left to the spontaneous activity of nature; with every rood of land brought into cultivation; every flowery waste of natural pasture plowed up, all quadrupeds or birds which are not domesticated for man's use exterminated as his rivals for food, every hedgerow or superfluous tree rooted out, and scarcely a place left where a wild shrub or flower could grow without being eradicated as a weed in the name of improved agriculture."

We hope to implore readers to take a little time out from the vast traffic all around us and consider several basic clues and canvases from history; manuscripts and ideals, legal, medical, scientific, economic, ethical and spiritual revelations that have in large degree shaped who we are and what we can do as a species, and as individuals. Those agencies of change – humanity as a whole, and individuals on their own time – offer up two compelling, but frequently contradictory journals of behavior. The variations can mean life or death to hundreds of millions of acres, and all the plants and animals therein.

Individuals are the ones who ultimately change species, accelerate evolution, make possible that which was thought to be impossible. A species cannot cuddle another – only individuals can do that. From a costs/benefits analysis, there is no greater opportunity in the 21st century than the fullest expressions of the human conscience. The 1912 essay entitled "Variability and Mutability" by Italian economist and statistician Corrado Gini has since been accorded much speculation under the guise of the Gini Index or Ratio; a measure of the inequality gaps between people, cities, cultures and whole nations, on a scale of 1-to-100, zero being tantamount to pure communism, or unequivocal equality. A generation of new questions must be asked in considering the nature of inequality. What about unequal distribution of pain, as opposed to pennies? To this terrible query we respond with what we outline under the penumbra of a "pain points index" and corresponding methodology for remediation.

"The Last Frontier"

This proposed metaphysic herewith enshrines nurturance and tolerance; unconditional love and compassion as hallmarks of the biological world. We acknowledge the reality but reject the necessity, as humans, of harming others. There are juxtaposed within this work multiple lines of inquiry whose intersecting nodes, examples, citations - and the prose in which it is all assembled - are intended to be indicative of critical initiatives we can engage as societies here and now.

This metaphysic is an orientation concerned, not with the salvation of souls, but the maintenance of ecosystems; of those myriad "hotspots," "coldspots," biochemical altruisms that speak their own languages in the infinite working out of so many biomes, the source of ongoing life and harvest and essential diversities. We take personally the maniacal assaults our species is leveling at the Earth, and so do millions of other people. This internalizing capacity is the last frontier, a cause for hope.

We propose science and ethics-based considerations, approaches to consensus building and the prioritizing of high biodiversity rarity, as well as so many other sullen, unheralded species that fall beneath the radar screens of near-extinction and are thus ignored by lawmakers.

This little light of ours seeks to honor certain remarkable personages, expeditions of discovery, astounding works of art and reflection. It is a most modest re-kindling of concerns and highlights that have advanced our love of the natural world in ways that cannot be denied or ignored, lest we decide one and all that we are so free as to have monopolized freedom for our own ends, thereby nullifying the countless choices that were born of necessity in a world of Others; of Nature that clearly values interdependency over selfish destruction, and rewards with ongoing life those sentient beings who have embraced the symbiosis of mutual fulfillment over the nihilism of self-involvement.

The ethical onus is upon this generation to restore, not to demolish, at every juncture where kindness and common-sense are not merely best practices, but the obvious choices to anyone who has the capacity to feel love in their hearts, and even the slightest inclination to make a change that helps others. Just this very day, in liberating scores of flies (*Musca domestica*) that had somehow been born within our California household during a period of extreme heat and humidity; taking special care with each and every one to relocate them out into the gardens of our home, we cannot begin to convey the joy we felt for their freedom. And in so doing, were reminded of the most obvious maxim: if a heavy, tall-standing mammal likely to live as much as a century; one who prides itself on some level of cognition and emotion is unable to save the most helpless insect, whose primary and ephemeral options (a lifespan of 15-30 days) lie within its compound eye of thousands of visual receptors, or ommatidia (which lacks a pupil and cannot focus), its wings, and ability to breed rapidly, then what hope is there for our kind? With 206-to-270 calcium-dense bones and nearly one hundred billion neurons in our respective brains, and weighing between us (Michael and Jane) roughly 4,480 ounces, to that of a single fly - 0.00035th of one ounce – where is our capacity to even begin if we had not taken it upon ourselves to spend a few moments saving flies? They could just as easily have been mountain gorillas, African elephants or other people.

Some meager portions of this manuscript have been adopted from the Authors' (Tobias & Morrison) work, **God's Country: The New Zealand Factor**, Zorba Press, A Dancing Star Foundation Book, 2011; as well as from Michael Tobias' Keynote Address at the Symposium on "Biodiversity Conservation and Animal Rights: Religious and Philosophical Perspectives," March 21st, 2012, at the University of London's School of Oriental and African Studies.

Michael Charles Tobias
Jane Gray Morrison

Dedicated To Josie, Stanley, Feather, and Charlie; to Mac, and To So Many Others We Will Always Love Unconditionally

1

COMPARATIVE METAPHYSICS

"The Creation of the World: And Then What?"

Art historian Campbell Dodgson (1867-1948) had been Keeper of (largely Old Master) Prints and Drawings at the British Museum from 1893-1932, specializing in Albrecht Dürer (1471-1528) and his age throughout the Benelux. He was also a theologian. In 1937 Dodgson donated, among other works, a very special piece of paper to his Alma Mater: Day Five of Jan Sadeler's "Creation of the World," an engraving executed sometime between 1580 and 1590 as part of a suite of the full Seven Days of the Creation. Day Five (British Museum #1937,0915.35) is particularly noteworthy in being the most complex choreography of avifauna and marine species (as well as mystery creatures) ever invented. In fact, there are probably more animals in this one engraving, done after a painting by the 16th century Antwerp master, Maarten de Vos, than in any other work of art, before or for centuries to come.

That would include all of the animals and intimated ones in Gerard De Jode's fine engravings during the late 16th century, done subsequent to Hans Bol's tale of "Noah's Arc"; or even in that remarkable image, p. 97, of Da. Sa. Bonaventure's biography in vulgate of St. Francis, published in Venice by Simon Galignani in 1593, in which Francis has a vision on a skiff, surrounded on lake and on land by approximately 106 creatures, including parrot bill and peacock, resting turkey, griffin and dodo.

Indeed, we know of no other painting of the Renaissance in which life is presented in such multitudes, with the exception of Albrecht Altdorfer's "The Battle of Alexander at Issus," of 1529, based upon stories handed down of the Anatolian combat of November 333 BC, which pitted Alexander the Great against Darius III, and their respective tens-of-thousands of troops. Modern historians suggest a number between 25,000 and 108,000 soldiers.

Of course, Altdorfer's magnificent sunburst upon the nexus of struggling life is biology in breakdown, human conflict; life that is doomed by its inner nature.

Whereas the tradition of the Creation, as embodied in the works of a Maarten de Vos and Jan Sadeler, seems to impart an ethical pertinacity. But that guarantee of survival does not extend to human beings, a fact Altdorfer seems perfectly at ease portraying.

Dodgson was a cousin of Lewis Carroll and clearly bore that extended family predilection for exquisite queries, phantasms, riddles and imponderables. The Sadeler etching invites contemplation of perhaps the greatest mystery ever rendered in such terms. The language of animals. There is God on the beach, having apparently "created" all that life around him – in the air, in the water, along the shoreline. There are at least fifty large creatures, from an ostrich and turkey, a turtle and a griffon, to many hundreds of smaller birds and marine species. Thousands, tens-of-thousands, are intimated. Their gesticulations are a verbal feast, proof of expansive cognition transcending all conventional paradigms. We see a world directly, absent the hedonism of human self-importance.

But what is to become of this sacred zoological fanfare, now that so many individuals of other species have clearly overpopulated the image? In fact, their exquisite abundance makes for a demographic crisis, a most taxing day of embryology, sea swells, imploding migrations, biological ideas convulsing after purpose. Should they fail to disperse, and quickly so, the piece of paper (height, 198 millimeters, width, 255 millimeters, or 7.79 by 8.85 inches) will be unable to hold them. In fact, can God himself – who stands scarcely any taller than the ostrich beside him – protect all this progeny? Is evolution (which a recent poll in the United States revealed to be disbelieved by at least 100 million Americans) going to somehow care for these marvelous creatures?

Or is nature to be left to her own designs, whatever they may be? This was the long-held assumption of Western science, enshrined in William Paley's 1802 **Natural Theology**. The question of longevity, of cause and effect, of so many astute gazes and high-octane testing of wings, of every biography represented in this vast panoply of personages, envelops the viewer in every possible dangling modifier. Charles Darwin broke through the conventions of Christianity's fear of asserting quantifiable fact in 1859 with the publication of his **On the Origin of Species**.

Neither insouciance nor nostalgia can abet this precarious moment in time. It is the true bearing of a compass compiled of metaphysics and question marks. Of some conjoined enigma, historically rich, wayward, beckoning, fully heralded with no guarantees, no lighthouse, no known methodology. Nor is there any talk of survival, or fitness; of robust genetic populations. Indeed, we see not one population but countless clusters of the same, or fantastically differentiated creatures. All of them, however, populate this invitation to speculate endlessly. There is no other prescribed set of calculations that can tutor our dumbfounded reception of so glorious and imperiled a scene.

Moreover, as is the case with so much art historical enigma, the 16th century context for Sadeler's masterpiece is no more evidentiary than today's contextualization of his benign genius within our own utterly out-of-control environmental debacles.

"The Word Become Flesh"

If John 1:14 is correct that – at least in Christianity – the word is God, then all that flying flesh and sea-foment illuminating every inch of Sadeler, is an embodiment of God. If God is Good, then God is meant to care for this assemblage of clearly confused and flailing beings.

But could there be some other cipher sending us back to the drawing board, to molecular biological, taxonomic revelations that would require centuries to even begin to work out?

Is the metaphysical injunction the appropriate school for learning, against so flabbergasting a natural history on the very day of its arrival? For René Descartes, no friend of animals, metaphysics referred to the taking up of space. In the case of matter, that space expanded indefinitely. For mind, it meant expansion into consciousness. For 20th century physicist Max Karl Ernst Ludwig Planck, the units of measurement of this space turned negative, namely, into the so called Planck quantum physics, for which Planck won the Nobel Prize in 1918, having done for sub-atomic particles what Einstein established for space and time. Planck recognized that there must be thought, consciousness, a god of some kind in all of these particles which, in themselves, however minute, contained the signature of a conscious entity, of moral balance between order and chaos.

One Planck length, or $1.61619926 \times 10^{-35}$ meters, is a constant. Its meaning bears little resemblance to any living being, or the space required of life. It is a pure doctrine, like gravity, which presents to us a living challenge: how do we grant balance and constancy of consciousness to the Creation?

It is a long journey from Jan Sadeler, who died in 1600, to Max Planck and Einstein. It is an even stranger journey by foot from God on that beach, to modern-day conservation and animal protection strategies. We continue to seek ways to protect all of those birds and fish, by every conceivable technique, from test tubes, stem cells, the manipulation of mitochondrial DNA, the creation of national parks and greenbelts and all of the calls to worship by environmental movements, generation after generation. The sanctuary movement inculcates Medieval British law, temples and tabernacles, people of faith and of no faith; of insects and coral reefs. Whether there is a god, no god, or any reason to be optimistic, the concept of extending the olive branch – first carried by a dove to Noah in the Bible, to let him know the flood was coming to an end - of hospice, no-kill zones, of taking sanctuary, is a profound testament to one's faith in the possibility of goodness. End of story. Beyond goodness and the olive branch, or in its complete absence, all bets are skewed.

But there is yet to emerge a viable corollary in the realm of protection to the delicate moment in Jan Sadeler's exquisite vision of God extending his hands towards the waters and the sky, giving loft and breath, life and death to all those he has somehow brought to bear in some fantastic scheme of things.

The protective spin-off is, at least in Sadeler, a matter of interpretation. It is not present in either the pictorial forensics or later iterations of the Creation. We are left to wonder: what is to become of all of these newborn creatures? Their consciousness, their souls, the composition of their interactions?

The word metaphysics derives from the Greek, meta (beyond), and physika (physics): Beyond the workings of God on that eternal beach (sharp crags on the horizon jutting from the sea), and the countless physical properties of Nature which we have come, in part, and by ever more curious yardsticks, to call by such names as Cavendish's Constant, the Compton wavelength and Schwarzschild radius.

Particle accelerators have gotten us no closer to the General Theory of Nature, to black holes, or to paradise.

Instead, we are smitten with natural causes and effects about, and to which we remain ignorant and exposed. Some tenebrous dark, teeming taxa of an Apocalypse; a biomass-like monster in and for itself, all headed towards oblivion, with ourselves, and all of our loved ones dragged down with it. More precisely, we lead the charge and it is a runaway train in the dark whose only apparent headlights have managed to highlight a surreal, at times deeply painful and poignant gallery of illustrious and hallucinogenic daydreams.

Mathematically, they – Nature's mysteries and impact on our own individual lives - add up to every God; every elaborate, enviable and elusive excursion throughout the ponderous valleys of our past; the maze of highlands towards which we seek the greener pasture. These are the biomes of our deliverance. They celebrate their life forms in and of themselves. They are purposeful. Of every distinction. Classically, their portraiture might encompass four sonorous streams cradling willowy banks, flowing from the paradise John Milton saw as a blind man; Ovid, Boethius and the ancient Chinese wilderness poet, Hsieh Ling-yun each imagined from strange and wearying exiles.

These hanging gardens of Babylon and Himalayan Shangri-la's catapulted Marco Polo, Ibn Batuta and Antonio de Guevara's early 16th century "Peaceable Kingdom wherein, as in the Prophet Ezeckiel's legacy, we "were in Eden, the garden of God..." (28:12-14), that concordium of Blessed Isles, Fields, mercies, ascensions, liturgies and St. John Chrysostom's locus of ultimate tranquility, as masterfully outlined in Jean Delumeau's classic study of the **History of Paradise** (1995). This would become Shelley's "Alastor, or, The Spirit of Solitude" located in a most fertile cocoon called Kashmir, on the edge of a cliff, surrounded by the forests in which one best commits suicide: the ultimate dialectic infecting our species' mixed motives, and sense of regret for what it has done to others; all those Others whose sentience exceeds and/or is different than that of our own.

In their many-tiered spectacle and distance from what is here, now, today, all these Arcadian evocations, perfections and paradoxical intimations of Being at the very zenith of all manner of resurrection form a most beguiling similitude within our collective and contradictory species which, endlessly, compulsively, has invented them for some peculiar reason. Note the perfect Adagio For Strings-like "Arcadian Songs And Dances/Sadness," (1948) by composer Virgil Thomson.

The rationale should be self evident. Yet, we have combed every frustration, plumbed each earnest daydream until – despairing of our endless circumambulations – have acknowledged the possibility, as did Samuel Beckett in his novella, **The Lost Ones** that these are nothing more than airy ziggurats that populate our most intoxicating myths; incite obsessive treasure hunts for humanity's genetic pre-romantic origins in unexplored limestone caves in the gorges of Gorongosa's Cheringoma Plateau, in today's Mozambique; or just beneath the summit of Mount Ararat, but cannot quite deliver on the promise we imagine once existed, like the library of Alexandria or Colossus of Rhodes; some Golden Age, as Plato envisioned.

This eternal return as Mircea Eliade thought of it, has not yet dampened the degree to which we seek a solution, some escape from the mathematical mirrors and labyrinths of the Dutchman Maurits Cornelis (M.C.) Escher. Our modern fixations, geopolitical, fitted with every clever device, are no less galvanized by so many glowing embers of hope amid dirty particulate matter raining down upon Greenland's reflective surfaces; the same embers and detritus of 900,000 year old caves and late night contemplations. For all of our rocket ships and nanotechnologies, we are still unable to communicate with a butterfly, a shark, or even another primate.

We have advanced myriad glitches yet scarcely a jot beyond Aristotle's **Nicomachean Ethics**, Maimonides' **The Guide for the Perplexed** (**מורה נבוכים**, Moreh Nevukhim) or Spinoza's "Deus sive Natura" in the Vatican manuscript of his **Ethica**. In Rembrandt's eyes, as Simon Schama reveled, a Dutchman had seen everything, revealed it to posterity, then died. Just like the wild tulip. What does that comparison mean? It is the core of the biological mystery that invites deliberations in accordance with the metaphysics, such as we are herewith proposing.

While it is tempting to simply accord a "seize the day" philosophy here and now, by way of a preemptory Coda, it does not satisfy all those endless legions of life we have betrayed. It wants to employ a deeper, more meaningful exit, certainly in the name of *Homo sapiens*. At the very least, we should be capable of conjuring a theory to explain ourselves that transcends the generous instinct of all that was the *aetas aurea* of the Renaissance.

"A General Theory of Nature"

A General Theory: one that explains the many endangered shark species as well as that bedraggled learned column of dogged pilgrims caught out in the wild peregrinations en route to Santa Catharina in the South-Central Sinai, all amounting to no more than some anonymous scaffolding of philosophy that builds upon its own Babels of Beyond, the Breughel family, Beethoven by the time of his 7th, Immanuel Kant at the height of his **Perpetual Peace**, Hesiod's practical farmer in **Works and Days**, and Heraclitus's **Fragments**.

This paradox of multiple personalities seeks out the 30-something Linnaeus, Buffon, Humboldt and Darwin – men in their prime - and a picture emerges rife with the combining potency of natural science and philosophy; of birds and poetry. There could be no more decisive rift in the chemistry of explosions: the perfect terza rima versus the absolute void (most palpable in Dante's **Divine Comedy**); paradise squandered, Arcadia reignited. This General Theory, in other words, is fraught with attitudes deriving from big personalities. Such is the more universally predictable method of science.

Consider the sumptuous chromolithographs in books like those of T. H. Fielding and J. Walton **During a Two Years' Residence among the Lakes**, printed in London for Ackermann, in 1821; C. W. Webber's **Wild Scenes And Song Birds**, published in New York in 1854 with incomparable color illustrations by Mrs. C. W. Webber and Alfred J. Miller, commencing with a Kentucky Mocking Bird perched above a divine landscape; or the unrivaled "Forest In Central Africa," one of numerous sublime color portraits within the most popular and massive work on African expeditions of the 19th century, namely, **The Life and Explorations of David Livingstone, L.L.S**., published in London by Adam & Co., in 1878. Add to these sterling observations, the extraordinary colored woodblocks defining the four seasons in the late 19th century Kyoto "4th Street" Maruyama-Shijō non-objectivist school**, Ehon – Yofu Gajo**, by Takuchi Seihô, Kikuchi Hôbrun and Taniguchi Kôkyô (1895). They all share in one common compulsion, a veritable aesthetic phenotype: the merging of art and natural history at a pulse rate and ponderation that is irreconcilably, irreversibly of the human animal. In Thomas Robins the Elder's Rococo gardens of the mid-18th century in Bath, or his formalist successor, the famed Humphry Repton's (1752–1818) numerous drawings and essays on landscape gardening, there is no doubting the human need of a beautiful picture in the mind to compensate for some enormous emptiness, a loss, or desperate hope to fill the gap. Yet, we remain our worst enemy in the larger picture, exorcists who would prefer midwifery.

This entanglement, so luminous a window on sapient dichotomies, is no more lustrously embedded than in the vast archives – some 9,000 works – of America's premier natural history philanthropist and educator, Graham Arader. Aside from possessing the greatest John James Audubon (1785-1851) collection in the world, Arader's works - by the likes of Mark Catesby (1683-1749), George Edwards (1694-1773), Francois Levaillant (1753-1824), Alexander Wilson (1766-1813), John Gould (1804-1881), Rene Primevere Lesson (1794-1849), Jean Theodore Descourtilz (1796-1855), Charles Lucien Bonaparte (1803-1857), John Cassin (1813-1869), and Johannes Gerardus Keulemans (1842-1912) to name but a few of the finest artist/ornithologists in post-Renaissance Europe, Australia and North America – have, to great degree, been made available to students at numerous universities throughout the United States. Says Arader poignantly, "My dream is to share this artistic and scientific legacy with our young and give them a reason to do the research that has been so invigorating for me for the last 42 years. There is nothing like the thrill of being face to face with the original work."

In assiduously striving to bring the works of art into the hands of students, Arader has been among that rarified few who have understood the primeval lessons, passed down from camp fire to camp fire: nature is what we are; its celebration and diligent study is our sole hope for survival. The notion of protecting nature is so ingrained in this impetus as to quality as a General Theory. If only it were.

"Cognitive Wilderness"

Throughout the 16th and 17th centuries in Western Europe, half-dozen families of engravers, including Sadeler, Wierx, and de Passe, embraced commissions to enlarge the visual scope of Biblical address, Noah's Ark, Adam and Eve naming the animals. Much of the 16th century business of drawing singularly pivoted upon Dürer, particularly such iconic subject matters as "Flight into Egypt," the "Small" and "Large" Passions (of Christ), St. Eustace, St. Jerome and St. Matthew. All prefiguring those last seven expressions of Christ on the Cross, "Father, forgive them, for they do not know what they do," in the Gospel of Luke 23:34. It was enshrined in a tonality we can understand by Joseph Haydn's "Die sieben letzten Worte unseres Erlösers am Kreuze," commissioned for a Good Friday service at the Holy Cave Oratory in Cádiz Spain and published in 1787.

Why the obsessions? Or, one thousand years earlier, similar preoccupations with what would soon be known as the Troubadours, the evolution of Provençal from Latin, sestinas of Arnaut Daniel, a great minstrel of nature who the historian Francis Hueffer, in 1878, reasoned would become the harbinger of ballads, blank-verse stanzas, rhyme-schemes of elaborate melancholy, all those "tell-tales" amounting to some compelling "fateful knot of passion" that would, in the end, invent Cervantes and his precise contemporary, Shakespeare, the penultimate metaphysicians.

The earliest French sonnets – "Does not, alas! all nature fade away, If from the fragile form the soul depart? I am that body – thou its better part- Where art thou?

–why this cruel, sad delay!" wrote La Belle Cordière, or Louise Labé, born in 1526 in Lyons, as translated by Louisa Stuart Costello in 1835. Lyons would, within a single generation, become the publishing capital of woodblocks illustrating the Bible, Ovid's **Metamorphoses**, and the vast majority of imagery upon which the entire Renaissance fell from brute naiveté to restless yearning.

By 1881, with the monumental publication of Jules Verne's **The Great Explorers of the Nineteenth Century**, philosophical import has given substance to specific geographical coordinates that have remained the core values of conservation biology, animal rights, deep ethology and mysticism, at heart. Verne concentrated his euphoric descriptions on journeys through Arabia, Elphinstone in Afghanistan, Guldenstaedt and Klaproth in the Caucasus, Raffles in Sumatra, the fate of Mungo Park, Mollien at the sources of the Senegal and Gambia, Cailliaud and Letorzee at the oasis of Siwâh.

Verne enticingly endowed the somber philosophies of Goethe and Dickens, Shelley and John Martin with topophilia, beloved places such as the Radak Archipelago, all those destinations which took up entire lives of French circumnavigators like Baron de Bougainville and Dumont d'Urville.

He chronicled John Ross's three Arctic expeditions – a Minister's son who had signed on as Apprentice to the Royal Navy at the age of 9; Verne profiled monkeys and "sea monsters," famed marble rocks, New Ireland, the Sultan of Madura and the Bay of Sharks. If he lived today, Verne would be the first to race to the center of that sudden gigantic hole at the "end of the world," Yamal, in Siberia, and probably – Cousteau-like – spend 31 days in its biologically-splendid Otherworld.

Verne was one of many who were stunned by the possibilities of seeking in a landscape all that epistemology and ethics can teach us. His fictions were nothing less than the elucidations of psychology; illuminated manuscripts of human nature. We have always been products, entire recreations of our environments and this has formed the original raison d'être of adventure and of introspection.

The quest for understanding what this conjoining of two ancient words actually means in one, metaphysics, and how it should come to be appropriately placed as a preamble and predicate for the act and substance of protection, began long before the word itself, which – over centuries – has become blurred in the outlines of Demos, the people of ancient Greece whose record of nonviolence is a mixed affair; and of meta-philology and the countless contrary nuances of natural science and its methods of understanding and expressing the world.

Not unlike Alexander Glazunov's incredible penchant for ambivalent reveries, as in his "Suite For String Quartet/Oriental," that musical countenance of all minor keys into some strange alluring beauty, metaphysical realities for Verne were a wonderful enigma of monstrosities, contradictions and irresistible expeditions of mind and body. This juxtaposing of nature with analytic phenomenology, as well as systematic rhapsody – two segments of the mind verging on catastrophe and incoherence, but sublimely so - was highlighted by Gottfried Wilhelm Leibniz in 1686 in his 60-page **Discours de métaphysique**, wherein he proposed the extraordinary concept that even whilst surrounded by evil, bloodshed, predation, every walk of horror, there was still a God who had chosen this world, this nature, as the best of all possible options; that very same garden Voltaire's Candide was well advised to cultivate at the conclusion of his life. For it was a certainty the day would come.

"The Hellenic Genius"

One good place to begin to comprehend this dualism of contradictory impulses that combine to form an ecological metaphysic is with the famed first century book collector, Apellicon of Telos (died c. 84 BC) who owned, among many others, original manuscripts by Aristotle. His enormous library placed him in peril and he would be forced to flee for his life, his private, Βιβλιοθήκη deemed by various Athenian authorities to represent a danger to the political status quo, in much the manner one might assume that Dmitri Shostakovich's music, particularly his Eighth Symphony of 1943 and Second Piano Trio (Op. 67) exemplifying the Jewish totentanz (Dies irae, or Dance of the Dead) would be perceived as a threat to Stalin's hegemony over the Soviet Union. Stalin evidently said "`Music illuminates a person and provides him with his last hope'; even Stalin, a butcher, knew that, and that was why he hated music," wrote the composer. [*1]When Apellicon died, his collection was gathered by Andronicus of Rhodes and then taken to Rome. Andronicus is credited with having established Aristotle's provenance guiding all future Aristotelian editions and translations. At some point in Aristotle's posthumous itinerary, Greek philosophical traditions had initiated a rousing chorus of debatable hypotheses pertaining to nature and human nature. With the tradition of Parmenides ('Being is; not-being is not') [*2] nature was subsumed by categories and "problems"; problems of substance, of impossible enterprise, material constitution, that which is beyond, or comes after, the physical, and the study of being as such. In other words, a chaos of extrinsic theoreticals with very little visceral or tangible connection to the world's children, and their parents, actually experienced, with their olive trees, seasons, birds and butterflies.

Parmenides was forced, by his philosophical musings, to renounce the mundane for the ultimate, and by suggesting the ultimate, the corresponding sorts of questions cascaded: what is the ultimate, if not now, in this very moment, as Zen philosophical traditions would many centuries later postulate? Is existence possible in the face of destruction, and if so, by what measurements? The Holocaust of World War II would certainly set the bar for destruction, and the meaning of nature in the context of human nature. The Holocaust would also challenge any and all precepts of philosophy. Whatever a human being might want, will, or imagine, the entire edifice of all that we take for granted could collapse in an instant, not as a result of fickle gods atop Mount Olympus but because of the sheer evil and insanity lurking in human nature.

The evolution of hypercarnivory, of mammals with a diet comprised of at least 70% flesh of other animals, affords a flimsy paleobiological explanation being an alleged need of more protein than that of hypocarnivores (consumers with less than 30% of their diet comprised of other vertebrates) and herbivores, and the changes over time in the ratio or volume of the post-carnassial (large, scissor-like) molars in various predatory versus non-predatory, species.

Such conjurations compound the difficulty of locating reality amid metaphysics, which no one has ever actually managed to encompass. We (the Authors) conjoin the noun with protection because, like the realms of protective umbrellas, metaphysics is vulnerable, connotes a fragility that is symptomatic of all ethical, ideological and biological change, fragility and sensitivity. Nothing about the nature of protection can be said to transcend the inescapable reality of our mortality and that of all other life forms. With that as a given, one can rationally assess the correlation between the biological sciences and philosophical and ethical thought, because they are joined, however one views the hierarchy of wants and needs, or calculates the numeric coefficients inherent to everything we perceive, believe, or hope.

"Phanerozoic Mysticism"

Hence, the "explosion of life at the beginning of the Phanerozoic and the accompanying rapid adoption of hard shells and skeletons." [*3] The Phanerozoic (comprising 542 million years – the Paleozoic, Mesozoic, and Cenozoic eras during which –as the Eon connotes in Greek, life appeared, or, "make life appear") is synonymous with all that would become the evolution of animal phyla and terrestrial plant life, some 3.5 billion years after the earliest known life forms on the planet. Biodiversity can be said to have fully proliferated across every planetary niche through the remarkable Phanerozoic cybernetics, or positive feedback, that engendered self-preserving niches which life, in all her diversity, would require. This nurturance, or system of safeties, is enshrined, hard-wired, both genetic and experiential. It is, in a word, protection, protegere, Late Latin, "cover in front." [*4] Hence, the Greek etymological underpinnings of what would become synonymous with a combining of the prerequisites for life and the very philosophical and physical realities accompanying life, A deep-sea octopus *Graneledone boreopacifica* protects her 160 or so gestating eggs for 4 ½ years, while the Mother starves to death. In terms of nurturance, the closest second protective exemplar of motherly love involves the basking shark, which can carry its young for nearly three years. An elephant's gestation is between 20 and 21 months.

For all three of these animals – octopus, shark and elephant - the operative phrase is "cover the front," then, not surprisingly, in conjunction with humans, across the landscape of our perpetual aggression, the apt riposte would be to "watch your back."

The protective "hard shells and skeletons" - in the 20th century – might first be characterized within the "wilderness" context by reference to the U.S. Department of Agriculture Regulation L-20 of the National Forest Manual of 1929, namely, "...a supplemental series of areas...to be known as primitive areas, and within which will be maintained primitive conditions of environment, transportation, habitation, and subsistence, with a view to conserving the value of such areas for purposes of public education and recreation." [*5]

When we consider the nature of protecting wilderness, we do not focus on anything like the atomic diameters whose distance varies from one to two Ångströms (1.0-to-2.0 x 10^{-10} meters), the wave-lengths of light emissions or number of molecules in any body of water. Nor do we consider the reality of cells, which, if magnified "10,000 billion times" would yield a vivid portrait of a strange, yet metaphorically complete reality: "To begin with," wrote Christian de Duve, "there is no such thing as the living cell. There are only living cells..." [*6]

Yet, as Erwin Schrödinger famously described, all organisms are dependent upon measures, rules and organizations of physical property. [*7] In looking at the focal point of any act of protection, the act itself – if truly pure- need not discriminate but, rather, exercise the full elegance of its impartiality by recognizing all substance as diffuse, distinct, recessive or dominant, entropic or progressive, unstable or stable, related to clockwork, or unrelated to anything ever imagined, invented, or discovered, measurable or non-measurable; in other words, an act of selflessness that would comport with every great spiritual notion of giving, of unconditional compassion, kindness, or hope. Some have even argued that our genes are optimism-biased.

From any cultural perspective, science is prone to overstate, or underestimate in its perpetual efforts to self-correct and challenge any and all assumptions. For example, in the case of hunter-gathering societies it was long assumed that such groups tolerated "little evolution, leaving slight chance for the adoption and spread of inventions." [*8] But that can hardly be true, since the world remains a hunter-gathering collective, its showcase of products having morphed into outrageous excess and tragic depletion. The evolutionary precursors for any physical inviolability now hinge upon human deliberation that can go well beyond the normal understanding of the commons – to encompass a compassion commons, footprint and unconditional speciation, self-correct for any and all limits so as to embrace a non-self organizing system.

Additionally, and to the core, this must draw our assumptions and conclusions towards a new storyline that assuages moral confusion whilst reckoning with inflationary demographic and psychological testimony. We are held hostage by this confusion, but also empowered to make changes – changes in the law, in our orientation to one another, and in the very constitutions that grant us some level of constancy within those countless and ever shifting communities we call home – one day *Sahelanthropus tchadensis*, another day *Homo habilis,* with an enlarged Broca's area in the brain, used typically for speech; then, by fewer than 150,000-200,00 years ago, *Homo sapiens.*

But whether in a cave outside modern-day Beijing, or on the Indonesian island of Flores, our traits, fire usage, teeth, tools footprints, spear-throwing, gracile frames, diminishing robustness, proclivity for art, are all tantamount to jousting for a granule of limelight, in the overall sway that "describes a waving tract," as the French naturalist Georges-Louis Leclerc, Comte de Buffon (1707–1788) once referred to a slow flying bird, taking its time, looking around.

"Wittgenstein's Tractatus Logico-Philosophicus"

Home is what it has always been, no matter the sheer panic of variability from place to place. Respighi's 'Ancient Airs And Dances/Book 1: Laura Soave' quietly elicits the music this global variety of Green Gods brings down to the enquiring mind. If the Earth is our evolutionary baseline, or – in modern methodology, the "control plot," then all true expression is actually inexpressible, however irrepressible its pastoral urgings. By that, we refer to the inexpressibility indoctrinated by Bertrand Russell in his May, 1922 Introduction to Ludwig Wittgenstein's **Tractatus Logico-Philosophicus** in which everything, said Russell, that is expressed must, necessarily, have no philosophical meaning because it is a tautologically impossible expression: to express something is to miss that which one would try to express, an endless series of indeterminate subtractions.

This is the same dilemma of expressing something in the here and now that wonderfully conditions the Japanese delight in ikebana, the Zen aesthetic informing flower arrangement, with its emptiness that is deliberately engendered to allow for ma, negative space allowing for positive egress, such as invisible breezes between flowers and stems.

This negative – as in photographic developing processes, as well as the inter-negative – connote our fundamental dilemma: Can we embrace the ideal of emptiness with any conviction when, all around us, the world infuses our sense with just the opposite? To posit the Zen, or the Dao, is to reflect upon counterpoints which inherently proffer a metaphysical space all around us, inside us, beckoning to this particular generation amid intergenerational misfits who must shoulder the burden of our failures, while sadly counting on we forebears to understand what is at stake here.

The fallacy of our cultures electing tyrants who behave with no faith in internegatives go on – as they do every day – to enact authoritarian contracts on our behalf that will, cumulatively, chip away at nature. From whatever vantage you may view this dilemma, it is, unambiguously, our degenerating problématique, whilst the human population surges with no surcease in sight.

Believing in that optimism gene rests on the imagined liberty to ignore fleeing seconds. But in the world of nature observation, sitting still for hours, days, to catch a glimpse of a snow leopard setting out from her high lair in a snowy cave in Mustang, or a Suriname Cock-of-the-Rock (the magnificent *Rupicola rupicola* at his lek (first documented by Robert Schomburgk in the British Guiana wilds of February, 1839, and which we first witnessed as a massive kapok tree collapsed and exploded not fifty feet from where we stood), poses the question: can we know anything about nature, given the physics of our inability to live for more than one second in any given second?

Wittgenstein was so focused upon the limitations of any language that it is not surprising he lacked any confidence in our ability as a species to actually talk about anything of substance. His 1922 edition was translated by a teenager who spoke a poor English, and Wittgenstein himself had to have felt the abusive contradictions inherent to family life and culture, having lost three brothers to suicide, and having as a classmate a very jealous young Adolph Hitler. Such facets of domestic back-story will indeed vary from tribe to tribe, village to village, family to family, as they have at least since the known emergence of human conscience as recognized by Ralph Solecki and his team during the excavation of nine skeletons at the Shanidar caves in the Zagros Mountains of Southern Kurdistan in Iraq (1957-1961). The Neanderthal skeletons themselves were dated to between 60,000-80,000 years old. It was the fourth cave, "Shanidar 4" that gives evidence of a true ritual funeral having occurred, with the use of numerous medicinal plants being analyzed through the remaining pollen samples: Ragwort, Cornflower, Joint Pine, Hollyhock and others.

Other grave offerings were discovered including pink calcite disc beads. Most of the bones from the Shanidar caves were scattered during the conflicts in Iraq and today the caves are threatened by the rock-fill Bekhme Dam at so called Great Zab 60 on the Greater Zab River, a tributary of the Tigris, near the village of Bekhme in the Sapna Valley of Iraqi Kurdistan. According to Nestorian Christian legend, this region – Mediterranean climate, invigorating montane open forest ecosystems rising to approximately 2000 meters – was the original Garden of Eden, with an ancient synagogue, three monasteries, and the dizzying array of spellbinding paleontological evidence unearthed by Solecki and team – possibly the most important archaeological finds of all time: the first ever recorded loving burial, as well as the earliest evidence for terrible forms of aggression between (modern) *Homo sapiens* and contemporary Neanderthal. [*9]

"I jump up: it would be much better if I could only stop thinking. Thoughts are the dullest things. Duller than flesh. They stretch out and there's no end to them and they leave a funny taste in the mouth. Then there are words, inside the thoughts, unfinished words.... and there's no end to it. It's worse than the rest because I feel responsible and have complicity in it..." [*10]

"Huis Clos"

Or, as Sartre wrote in his play **Closed Doors** (**Huis Clos**), "à bout de souffle", out of breath; [*11] In his Introduction to **Nausea**, Hayden Carruth adds that "Existentialism is opposed to the entire rationalist tradition deriving from the Renaissance…." [*12] Throughout Sartre's earlier **L'être et le néant**, translated as **Being and Nothingness –A Phenomenological Essay On Ontology** [*13] he would iterate that "there will appear a self-consciousness in general which is recognized in other self-consciousnesses and which is identical with them and with itself." [*14] Philosopher E. M. Cioran, in a similar vein, has spoken of "a moment" "in the life of the mind" "when style, transforming itself into an autonomous principle, becomes fate." [*15] We perceive the acoustical equivalency in such works as Sergei Rachmaninoff's "Vocalise Op. 34 No.14" (1912, the last of his "Fourteen Songs") and Ralph Vaughan Williams' "Five Variants Of Dives And Lazarus," Fantasia Tallis/The Lark Ascending." (First performed by the New York Philharmonic at New York's Carnegie Hall June 10, 1939, conducted by Sir Adrian Boult).

What does it all mean, these anomalous citations; this list of abbreviated incidents in the history of the mind, one mind, a mind? DNA originated in conditions of crystal mineral protective structures, according to organic chemist Alexander Graham Cairns-Smith. His "clay hypothesis" of abiogenesis or biopoiesis (the genetic locus replication whereby the earliest DNA sought out microscopic clay structures equivalent in literal form to that of the double helix) provides ample visual cortex residua bringing the evolution of consciousness back to its first veritable spark in the primordial soup. Human DNA converges in the primatology of a nearly singular attitude and orientation whose genetic distinctions from bananas, pocket mice, sea slugs, even Precambrian stomatolites, is only barely differentiated in degree, not kind. All that distinguishes the neurologically skilled marine Gastropods, including the sea slug, from its terrestrial cousin, the many land snails, is its manner of occupation: a one, two or three-bedroom home (a shell), or no shell; a home, in other words, encompassing its very surroundings. This latter distinction ushers in a pantheism outside the clay hypothesis of Cairns-Smith, arguing for the home of every wild child, whether a Tasaday cave dweller, or the former Caucasian pygmies of Arunachal Pradesh, as reported upon by Kingdon-Ward in the 1920s.

Are these peopled architectures of great importance? Probably not. While paleoanthropologists make much of the anatomical changes of our forebears, particularly with respect to the evolving size of our cerebellum, as measured in cubic centimeters, the reality of today's un-contacted tribes defies any exceptionalism that might be deployed to bolster humanity's claim to something new under the sun.

The evidence for distinctions is equally slim when considering fellow prosimians and simians (all the monkeys and apes, tarsiers, galagos, lorises and lemurs). The reasons for this are quite systematic: Regardless of the so called Eoarchean catalysis in chemical systems (that initial "spark" of life over four billion years ago) we are all philosophers, and among them today can be found a generosity of spirit – old Taoist texts, Jain canonical doctrine, Buddhist and Shugendō revelations, as well as our close hominid cousins, those happily bathing Japanese Macaques at the Jigokudani Hot Spring. Such converging genetic and orienteering rivers of being attribute equally a philosophical bent to every living creature, which goes to amplify the point; and if not philosophy, then certainly the respect of philosophers, for the inchoate life force that animates all sentience without correlating for morphology, adjusting for physiology or need, or nitpicking with regard to the size of the brain case or the number of cerebral neurons.

Most of the primate communities rarely exceed 150 individuals, but baboons and some of the Geladas can number (at least temporarily) over 600 members of a "herd". [*16]

This leaves the notion of conservation, of ecology, of compassion and biodiversity as dangling modifiers in a long tradition without definition. Our ancestors comingled sexually; they successfully procreated, as clotting and red blood cell data among Tibetans has recently revealed. Differing Homo species mated as recently as 25,000 years ago, if not more recently, as discovered in the Denisovan EPAS1 gene.

"The Evolution of Critical Uncertainties"

Just as natural selection has unequally distributed the advantages of sexual dimorphism, female choices, and the role of Alpha males and females during courtship as well as all subsequent child-rearing orientations, of even greater importance is the fact that the evolution of consciousness does differ on critical turning points from that of the actual conscience. Circumstances dictating nurture over nature cannot be denied. What is more difficult to assess are the outcome modifications from apperception, self-awareness in the world and the prospects emanating from that internality of compassion, whether of the same clade, con-specific or generalized.

In other words, not unlike others, *Homo sapiens sapiens* have evolved to varying degrees conceptually and ethically over the course of the previous tens-of-thousands of years. The last Neanderthal may well have been vegetarians in the Near East, as well as in the pre-Dynastic and Old Dynastic Nile River Valleys of Egypt. A non-violent cultural paradigm overpowered during the Middle Kingdom by the aggression of meat-eating *Homo sapiens*, who co-habited the same geographies. That there has been ethical differentiation within the same genus umbrella of Homo is genetically undeniable. The question remains, then: could Saddam Hussein have successfully mated with Mother Teresa? Or were there ethical delineations (*incertae sedis*, a level of uncertainty) sufficiently broad to have connoted actually taxonomic drift, the equivalent of subspecies? Natural selection is a powerful tool, particularly in light of the fact that different species entire have mated successfully, at least for one generation's worth of hybridized offspring.

"If each pair of Blue Tits or Black-capped Chicadees produced eight young in a season and all survived, there would be five pairs breeding the following season. Continue this process over ten years and the population has multiplied almost ten million times!" [*17] This provides the fuel for genetic experimentation. Such numbers games also escalate partnerships, symbiosis and bridges (the bridge being an iconic symbol for the Club of Budapest, as well as connoting those pathways Marc Bekoff describes in his brilliant and accessible thesis regarding "rewilding the heart").

In the case of the *Quelea quelea* (Red-billed Quelea) – considered to be sub-Saharan Africa's equivalent of the once most multitudinous North American bird, the (now extinct) Passenger Pigeon, Queleas number between 1.5 and 5 billion breeding pairs – making it the most numerous vertebrate, by far, in the world, especially given its life-span of two years, and hence a turnover rate of individuals whose only more numerous competitor can be found amongst the various social insects, like super-colonies of the red fire ant, *Solenopsis invicta,* spreading across a network of hundreds of miles, or lice, chiggers and mosquitoes.

Interestingly, unlike other species containing large numbers of individuals, mutations among the quelea are little understood, if at all deciphered. Whereas, in the instance of the aformentioned changes in humankind – there is no doubt. Writer Charles C. Mann in his essay, "State of the Species – Does success spell doom for *Homo sapiens*?" [*18] "... most researchers believe that in that flicker of time, favorable mutations swept through our species, transforming anatomically modern humans into behaviorally modern humans." Mann suggests it happened in as short a time as 50,000 years.

"Indeterminate Obligations"

While it is a general belief that mothers nurture their children, not all of them do. Nor do all fathers feel the same level of devotion, or even knowledge of their offspring. A male moose wanders off into a life of the solitaire. A father of 58 cannot even know, or recognize some of his children. They stand before him on holidays and announce themselves, and the ritual becomes increasingly awkward and amusing as they grow up, some die, others leave the country for another, and – in the end- the family ties are without anything like the fidelity that humanity would ordinarily attribute to a close-knit bond defined by kin. It is said that the Sharifian Emperor of Morocco, Mulai Ismail (1646-1727) had "342 daughters" and between "525" and "700 sons. **The Guinness Book of World Records**, 1997 edition, records "the wife of Feodor Vassilyev, a peasant from Shuya (near Moscow)" who "gave birth to a total of 69 children"...[*19]

Such family ties are hard to reconcile with the other extreme, that alabaster tomb constructed in 1419 for Sir Ralph and Katherine Green at Lowick, Northamptonshire, and other tender couples immortalized in their hand-holding monuments. [*20]

Add to this ever expanding complexity that pits unity and the intimate against demographic odds that are, increasingly, without boundary and the human condition drifts from unity to discord. New United Nations data suggests that, on the high side, should the human species even survive the coming decades, by 2,300 our population could hit "31 billion". [*21]

With the conservation conundrum now enshrining our hopes of survival, the metaphysical rudiments of our emotional ability to withstand the assaults we, and the world we inhabit have unleashed become the sole focus of all philosophy, psychology and biology, though there are few mechanisms for understanding what is happening to us.

"...with regard to the indeterminate obligations due from us to others," wrote Immaunel Kant, "... presents us with the idea of the will of every Intelligent as UNIVERSALLY LEGISLATIVE." [sic] [*22]

The fact that Kant predicated this universality on his bias in favor of humanity coincides with the practical reality that there is no animal rights if there is no humanity to propagate animal rights, realizing – all of us – that animals do not need us; we need them. Kant, intuitively, grasped that because humans exist, their very existence poses the threat that will not be stayed, obviated, or ignored as long as we are here. That said, conservation, animal rights, amelioration of suffering of others will come about as a result of our intervention, our humanity. By this reversal of priorities – otherwise counter to any rational animal rights movement- Kant has actually elevated the reality of any interspecies idealism to its appropriate place. The burden is upon us, suggesting some kind of sublime predicament of humanism. [*23]

Sartre, in his book entitled **Existentialism**, contested humanism citing Cocteau's ridiculous notion of a self-importance (odd, because all things Paris impart convivial-ity, distributing ego to both Left and Right Banks without bias) by way of the invention of airplanes in his "Around the World in Eighty Hours." In fact, said Sartre, not even a "dog or the horse" would follow for the universality of airplanes. They want none of it; feel no pride in the fact of their existence, and have more sense than we do as a community of species. [*24]

Voltaire spoke of the "Religion of Nature." [*25] Sartre backs away from the position of religion, embracing instead the protective truth of every mother towards her child, which he describes as "natural condition...the prime reality." [*26] Here again we revert to the moment of birth, when an offspring is imprinted at once upon his/her/its progenitor's guise, if not the very eyes.

The community of dogs and horses must be viewed against the other community, of mothers and their children; of whole human populations hiding behind walls or within the isolated clusters of monastic life that Boccaccio lent medical and political urgencies to, by way of irony. Irony, in turn, is the cumulative fortress we have erected to help accelerate our disassociation – like some Icelandic tectonic plate – from our Paleolithic past, where unison, and a universal language of nature was easily accessed. These intimate, fear-driven layers of our early history have few analogies today, unless one were to re-interpret megacity life as an architecturally-transformative possibility, though this is a case for urban planners who may have little or no dialogue with the trans-species contests that rule the remaining biological norms of life in the 21st century. Our current state of magazines are rife with re-engineering, terra-forming rooftops in Lagos – with its 15-to-20 million souls, soon to double in number, 80% of whom will be children inheriting a family budget of approximately (US) 1 dollar per day.

"I-Thou"

For Martin Buber, hailed as one of the great voices for the renewal of faith in God and inner dialogue with God [*27] the notion of the Other, of trees and species of all kinds being protected, loses all significance because, According to Buber (who – in analyzing the nature of a "tree" – wherein his "relation" with it "is mutual,") he reinforces the complex biological interconnectedness of all beings but concludes (in the confused discourse dependent upon the anthropomorphized "He" and "She") that of the tree's alleged consciousness, all those connections to the soil, the wind and the stars that is the tree, he supposes, "The tree will have a consciousness, then, similar to our own? Of that I have no experience… I encounter no soul or dryad of the tree, but the tree itself." [*28]

Ich und Du ("I-Thou" or "I-You"), the book first published in 1923, represents in sum a relationship and as such provides for every level of qualification. While his emphasis was upon Thou, not It, which he stripped of all sacredness, his ambiguities are so stacked in favor of the entire Creation that Buber is indisputably in favor of trees and flowers and cats. He cannot dissociate himself from that which his God has favored and scented the Earth with. As one species among so many, even an old philosopher – carried on the shoulders of Israeli military during a bombing, refusing to leave his study – in the end becomes one with God by smelling the roses, not turning away from them.

Christopher Chapple of Loyola Marymount University has rendered a lush translation of an eleventh century Kashmiri Sanskrit version of **Vasistha's Yoga** (**Yogavasistha**), by Valmiki (also the author of the famed epic **Ramayana**). Valmiki was a 5th century BCE Sanskrit poet and Yogic master and the **Yogavasistha** consists of nearly 32,000 verses. As Chapple describes it, "the author sets forth elegant descriptions of progressive meditations on the elements, leading to a sublime state of connectivity." In verse 89.58 he says, "Through performing concentration on the earth, I dissolved into the form of the earth...." Then, in (VII.92.1) Valmiki declares, "Next I came to concentrate on the operations of the wind, spreading my thoughts resolutely to examine the world and satisfy my curiosity. I rose up into the wind [that surrounds] the earth, playing with the people, the trees, and the beautiful flowers, protecting the water lilies.... [The wind drives] the hard showers of rain and snow...For the gentle ones, it brings silence and innocence."

This connectivity relates implicitly to the whole nature of protections in conservation biology, but also in every aspect of interspecies relations. For Jewish philosopher Martin Buber, this similar essence could be achieved in the Ich-Du (I-Thou) relationship (Tat Tvam Asi in Sanskrit –Thou Art That, as Chapple characterizes what is one of the most famous of the Mahāvākyas, Grand Pronouncements, in all of the Vedantic Dharmic literature, originating in the **Chandogya Upanishad** 6.8.7.)

With many humans withdrawing into a kind of silence, kindness itself must discover new equations that can muster some level of idealism where pragmatism is not simply the quixotry of past visions that died out, or of invented concepts, like Buber's "faith" in which prayer is vested only in that personal relationship between an invented human conscience and its God. There is, beneath these heavy philosophical stakes, a more easy-going, if less dramatic, bottom-line that can assuredly lead even the most heartfelt skeptic into a gentler, less demanding circle of friends wherein a practical ethic may suffice to lend what humanity we have left in us to all others, unstinting, unobjectionable, and without surcease.

"The Vagaries of 'Protection'"

Perhaps this practical course of ethics was never more rationally spelled out than in Mrs. Trimmer's remarkable **History Of The Robins – For the Instruction Of Children On Their Treatment Of Animals, With Twenty-Four Illustrations From Drawings by Harrison Weir**, in which she writes with astonishing clarity, "...it is really a meanness to destroy any creature because it is little, and in children particularly absurd to do so; for, upon this principle, they must themselves expect to be constantly ill-treated, though no animal stands more in need of tenderness than they do for many years from the time of their coming into the world; and even men and women might expect to be annihilated by the power of the great Creator, if everything that is little were to be destroyed." [*29]

This sentiment could not be more tragically underscored than in the case of a New York mother found guilty of torturing animals to death in front of her children. She received a prison sentence considered the "maximum" for her crimes – of two years - for keeping what was described as "a concentration camp for animals".

Conversely, children have a strikingly different example of the adult human psyche's relation to animals as the late Seamus Heaney wrote in his first published collection of poems, **Death of a Naturalist**, "When Francis preached love to the birds/ They listened, fluttered, throttled up/Into the blue like a flock of words...Danced on the wing, for sheer joy played/And sang, like images took flight. Which was the best poem Francis made..." [*30]

But, to recall, Francis, and the naturalist in Heaney, exist in contradistinction with humanity's ever-escalating assaults on the life of birds. One study, among thousands, reporting clearly on the complete aversion amongst a sample of 59 avifauna species, of 28 percent forgoing their critical migratory stops when auditory signals of human road traffic were broadcast; inflicting extreme the "ruckus" peril upon the normal needs of "robins, sparrows, finches, chickadees, and thrushes," but particularly "Cedar waxwings and yellow warblers." [*31]

Such audio pollution is also reckoned to kill at least 300,000 people per year through stress-induced complications. But then, when we add illimitable habitat stress and fragmentation that clearly inflicts the most mathematically far-reaching horrors for biodiversity, in addition to direct and indirect animal cruelties and human consumption of wildlife, all under the exacerbating umbrella of climate change and the release of carbon and methane from the soil and deep oceans furthering the chemical and meteorological crises (20 tons of CO_2e in the U.S. per person versus an international average of 4 tons). This end-loser summation of the Anthropocene represents a global catastrophe equivalent to an extinction level event.

What cannot be overstated is the potential for a rare event to occur, something unprecedented. In mathematical and stochastic (the predictable evolution of random values over any proximate duration) terms this could mean a number of dreaded comparisons for biodiversity; for humanity's cruelty and ecological footprint: a Markov blanket or chain, named after the Russian mathematician, Andrey Andreyevich Markov or Markoff (1856-1922). Stochastic Process Rare Event Sampling is a computer algorithm for simulating various probabilities. Markoff processes, entities, paths, chains and blankets consider the likelihood of rare events occurring if we lack information pertaining to the future and past and must resort to deducing our conclusions based upon real-time current situations. Ecological crisis that is unprecedented, and whose future can only be imagined, falls into this heady if hideous category: we cannot protect that which falls outside of our ability to identify or study from past catastrophes. It means our extrapolations are useless. We can only focus on today, not the "Hot Zone" calamities drawn-up in the guise of Ebola Virus symptoms (human faces essentially coming unglued from bone structure and other unthinkable scenarios), but the entirely precision-loaded data from the current extinction literature and the all-too-real truths behind the closed doors of slaughterhouses.

While mega-losses occur in boom and bust species periodically – whether Arctic lemmings every 3-4 years, or Pacific sardines whose oscillating populations can occur over 50, 60 or more years – these drastic and sustained crashes in populations are now in direct sync with such unnatural impacts as over-fishing in combination with changing oceanic temperatures, leading to negative trophic cascades. Both of these species, lemmings and sardines, are rich in nutrients and countless other species depend on them for survival, from Arctic Foxes and Snowy Owls to Brown Pelicans and various seals. Now, all are imperiled by starvation, not just the individuals, but by natural extension, breeding dens, litter sizes, indeed, entire populations. Perhaps we cannot control the Pacific Decadal Oscillation, affecting the cycle of nutrient turnover in the oceans, which can alter the rate, temperature and movement of deep cold waters as they move towards the surface, in turn, naturally affecting food chains and keystone species. But humans are not dependent upon sardines to survive.

Hence, it is obvious, as numerous environmental groups and researchers have attested for many decades, that complete "no-kill zone" mandates must be put in place to at least eliminate those deadly oppositions in nature that we can control, our own consumer/industry choices. These are the front-trenches that summon our basal concept of "protection," Aegis (Αιγίς) in ancient Greek, losely thought of as some kind of magical pouch from Libya. A goatskin coat; a shield; a hurricane driving away the demons. A gorgon combined with the skin of a huge snake guarding the Golden Fleece; a dragon (Druk) who shelters all of Bhutan; a Raven (the Bhutanese national bird) who once protected a King and his entourage; a 4th century BC Nubian image of Isis, uncovered from Sudan; a Norse dragon; Old English beorgan; Middle English proteccio (as of around 1275); a shield in 12th century French; prōtēctiōn in Late Latin (a colloquial emendation to the aforementioned "cover the front").

By 1880 "protection" was applied legally to birds in Britain (although, not entirely so – a clear beacon on the inherent increments of political jargon that has obfuscated singular discernment and resolution in nearly every law, treaty, alliance and judgment handed down in every environmental piece of legislation, or hearing since the earliest documented court case involving a non-human species, namely, the proceedings attendant upon the sadly execution of a pig in 1266 at Fontenay-aux-Roses. In 1750, a proud jenny (female donkey) was acquitted in a court where witnesses attested to her virtue. The human accusers, however, were treated with astonishing ferocity. It was a lesson barely heard by subsequent law students or judges, though E.P. Evans' **The Criminal Prosecution and Capital Punishment of Animals** (London, William Heinemann, 1906) and Sadakat Kadri's **The Trial: Four Thousand Years of Courtroom Drama** (New Random House, New York, 2006) embody a swathe of legal anomalies that would – in any species-egalitarian society – evince the requisite resolve that takes its cue from earlier briefs and important interspecies precedents.

In 1889 the Plumage League was formed in Manchester by two women, Emily Williamson, a philanthropist, and Eliza Phillips. It would gain "Royal" status in 1904, having merged with the Croydon Fur and Feather League. Together, banning the use of plumes and leather, or the harming of any birds – except for food, and excepting, oddly, the ostrich - the organization would soon be known as The Royal Society for the Protection of Birds. A century later, it would morph into the "Wildlife and Countryside Act."

A dog, "Rupee" rescued from a garbage dump in India, moments from death, brought back to life by a loving human companion, the two of them eventually managing to trek together to the base camp of Mount Everest (November, 2013). The photograph from Rupee's triumph speaks volumes to this potent possibility that is kindness amongst species and individuals. [*32]

As expressly, protection also embodies the basic conservation principle of "Kubu," which in Bahasa Indonesia (Indonesian) means "stronghold" and in the case of the Kubu Village Forest in Central Kalimantan, where Orangutan Foundation International ("OFI") works, among other regions, it means protecting all one million acres of the Tanjung Puting National Park, the 188,000-acre Lamandau Reserve and the 1,000 acre Pasir Panjang Village Forest, the location for OFI's "soft-release" facility for rescued orangutans. As described by OFI President, famed primatologist, animal rights activist Biruté Mary Galdikas, "This will create a vast, province-wide system of contiguous, biologically intact nature reserves, parks, corridors, and protected areas, which together will help provide essential habitat to ensure the long-term conservation of the endangered orangutans in our region." Explicitly, that means providing "for patrol posts and water towers to help protect this vast forest region from illegal logging, mining, hunting and forest fires that may sweep the region during the dry season." [*33]

An "individual" can be afforded rights and a long history of advocacy devolves to that personhood within every organism, varying, of course, from culture and ethical tradition to another; from absolute venality to thorough veneration. Human nature is steeped in the love of unpredictable contradiction and every version of anti-social personality disorder affecting all other life forms. We continue to debate rights, duties, obligations but can find no pellucid or enforceable standard with respect to individuals that are not human. Those surprise others are caught out in the whirlwinds of our complacencies, neuroses and unprecedented capacity for inflicting pain and biological chaos.

Historically, then, the word "protection" has evolved with a broad set of meanings that are hard-pressed to account for human nature. Choose your potion, select your gravestone. Self-protection, no first strike, "specially protected species," and continuing uncertainty as to what constitutes the first truly "protected" area, or individual, or population, in the world steadfastly shrugs off the notion of "No Kill." We are, in other words, activists campaigning in defense of our insanity. No electron in the furthest-most galaxy from Earth was ever so lame-brained. No particular systems theology – in physics, electromagnetism or quantum mechanics has the language or tools to account for this brazen and self-elected illness.

Epping Forest? A Pharoah's tomb? A sacred Paleolithic amulet? Against this confusion, and in any ecological context, what can the U.S. Declaration of Independence possibly mean? Tort law would have us believe that the "protection of life, limb and freedom of movement" were universally subscribed to in all legal systems. That protection from "intentional interference," or "unlawful force" or "personal injury" were written into our genes. Nothing could be less clear or certain. And when these noble standings are suggested for other species, "animal welfare" is proposed as the generic "minimum" standard upon which all codes of bioethics and animal ethics committees take their cue, as we detailed in our exhaustive study of New Zealand's animal ethics committees and their influence upon the final judgments handed down across paradise.

"Minimum standards". That is the best our protective measures have thus far been willing to concede nature and her offspring.

Today, we have any number of "units" of protection in every conceivably designated legal or customary practice, usage and specificity. For example, the Data Basin Protected Areas Center (PAC) uses national and global geospatial boundary information to assess and consolidate critical terrestrial and marine information within standardized GIS and IUCN coded formats. [*34] But we have no uniformity whatsoever when it comes to pain, cruelty, individuals versus populations, Families and Genera, species and sub-species, clades and races, sympatric (different species co-existing in the same geographic region) and allopatric (species geographically isolated from one another) individuals. We know that countless sub-species, possibly differing species, can mate. Whether their offspring are successful, or not, is a matter of natural selection which, given the vast body of data suggesting mutualism and co-dependency, as well as basic symbiosis, there is every reason to believe that speciation is as culturally driven as it is individually shaped and lived.

Speciation is a term first applied by the American botanist Orator Fuller Cook (1867-1949) to differentiate "the splitting of lineages or 'cladogenesis,' as opposed to 'anagenesis' or 'phyletic evolution' occurring within lineages" [*35]. Molecular systematics, providing tools to assess deep lineages, or ancestral lines within evolution, commonly thought of as a tree of life, a coral reef of communities, each successor generation taking its direct genetic influence from a predecessor, is no guarantee that classic punctuation marks in evolution will not bless some altogether unexpected mutation that thrives; or engender safe harborage for what might otherwise be deemed an unsustainable experiment; a living theorem that died out; a concept that collapsed under the weight of its own untenable genes.

Life is rich with serendipitous ambiguities, as is death, particularly when one follows cells after the death of an organism. They do not all die; and some creatures have been discovered that continually regenerate themselves, thus falling into a category for which "immortal" is not a reach. *Turritopsis nutricula* is one such long-lived jellyfish, possessed of a tenacious DNA-protecting ribonucleoprotein, the celebrity enzyme believed to help reverse the aging process, telomerase.

The debates across the life sciences are not about the absolute validity of evolution (let alone, some higher calling entitled Creationism – a lyric, a lovely sounding mythology, one that Ovid built his 15 transformative books of the **Metamorphoses** upon, from the first Six Days of the World, to the Godhead that Julius Caesar was to become). Rather, those intellectual firestorms and self-correcting ardors concern the genetic varieties of experience, and how such multiples of experience and perception are bred into the vast storybook that is life on Earth.

In broader terms, solid scientific research, by definition wired for self-doubt, harbor the healthy fundaments for everyday conceptualizations of some "new nature(s)" that may be any one individual's unique birthright. By that denomination we refer to an individual of any species, whether sea squirt or the phenotypic plasticity of aphids, carp, or Speckled Wood butterflies; creatures that can rapidly change under stressful environmental conditions, particularly climate change.

That said, and this is the most disturbing effect of all: the accelerating Anthropocenic implications for convergent evolution, whereby ancestral clades and their contemporary descendents form commensal behavioural strategies and concentrated phylogenetic compensatory manifestations, is not occurring by any empirical macroscope. Indeed, data recently revealed suggests that as much as half of all known invertebrates (insects and spiders) on Earth have vanished as the human population has doubled. (*36)

"Protecting Phylogeny, Evolution and Biology"

In a recent analysis of Mexican bird species, a set of questions was posed, methodological approaches to understanding true taxa differentiation that had not been focused upon since the time of Darwin – and all but ignored during the past 150 years. Adolfo G. Navarro-Sigüenza and A. Townsend Peterson examined biological, evolutionary and phylogenetic species concepts, revealing debates within various scientific camps that easily change the nature of what is considered a distinct species. "A total of 135 biological species was divided to produce a total of 323 phylogenetic/evolutionary species, 122 of which represent "new" endemic forms in Mexico." [*37] Policy implications for such re-assessments are enormous. [*38] Every nation and biome contains the sum total of scientific standards, arrived at post-1758 or so, the year of Linnaeus' 10th edition of the **Systema Naturae**, with its 4,400 recognized animal species, and 7,700 plants species, as well as all the major revisions to his binomial nomenclature approach to delineating organisms. In just one country,

Mexico, the following conclusions were reached if phylogenetic and evolutionary species definitions were to be optimized with respect to enacting new legislation to enrich true protection for that country's birdlife: "The current official Mexican endangered species list, BSC based, encompasses 371 birds, ranked as 277 species and 94 subspecies. The same list of protected forms changes under the phylogenetic species concept because 47 of them are not recognized as valid species, while another 28 forms merit higher levels of protection. Additionally, under this concept another 11 forms should be candidates for inclusion based on their restricted distribution. We call attention to the fact that the use of one or another species concept affects endangered species lists." [*39]

In fact, as speciation gains traction and more and more organisms are identified as unique, the enactment of commensurate legislation falls father behind. With human economic supremacy defining all turf, whether within borders or international waters, the legalese is lost in antiquated, obsolete, or impossibly competitive advantages wherein the life forms at stake are simply dismissed. As defined in terms of economic advantage, the field of conservation is increasingly at a distinct disadvantage with each new discovery, at every biological threshold. Whether the stones are turned or unturned, the harm meted out is only increasing.

"From Linnaeus to Shinto"

Interestingly, a convergence of spirituality and evolutionary theory can easily be read into certain protective deities. The Shinto Sun Goddess Aka Taiyōsama Amaterasu, for example, one of some eight million+ largely female deities, is the most critical Japanese Kami whose residency at the base of the ancestral spirits are the Musubi or interdependent, transgenerational natural entities of the Universe – Energy, free spirits willingly enshrined in waterfalls, caves, trees, every conceivable animal - that protect, preserve, and are in harmony. All of her descendants are of the same clan [read: clade, or phylogenetic tree].

Whereas the vast majority of ethnographic literature traces matrilineal or patrilineal descents, Shinto vastly opens up the relatedness of all beings. Anyone, Shinto, non-Shinto can make pilgrimage, or Omairi to any of the 80,000 Shinto shrines formally acknowledged within the Shinto Association Jinja Honchō that watches over these spirit (kami) inhabited places. Children traditionally are given their names after one month of life based upon the name of the closest local kami, who protect that child forever.

Many of the Japanese shrines are guarded by a pair of magical lion-dogs, Koma-inu or Shishi. Popular protective spirits, carved in stone after such historic figures as Jizō Bosatsu (circa BC 500) who protect travelers. [*40] But the list of protective deities in Japan, India, indeed in every region of world culture defies enumeration in this venue. It is sufficient enough to hint at the variety, from Gaia and Demeter to the Tibetan Buddhist Six Armed Mahakala; from the Egyptian God of Bes, meant by his long beard to scare away demons, to the jinn's of ancient Nabataean Palmyra. [*41]

"Help Me"

In every language, 'help me' – aidez-moi, pomogite mne, qing bang ge mang, watashi o tetsdatte kuremaska, Bitte -hilf mir, Aiutami – incites the quintessential foundations of any metaphysic. From inward-dwelling queries of every dimension, state-of-mind, attitude, personal weather, long bleak or fleeting ecstatic reaches of one's dreaming qualifiers, to real-time protection on the ground, these are all corollaries of the same injunction, namely, an appeal to the heart that can be translated into safe harborage in one's own home, backyard, community. A global and contiguous tapestry of welcoming sanctuaries, no different than Operation Damayan, "help" in the (Filipino) language of those hundreds-of-thousands of survivors, those injured and overwhelmed by Typhoon Haiyan.

Metaphysics also conjures up the myriad of complicated prospects in a world where triage butts heads with unconditional love, pragmatic idealism, and the solace of truths regained where paradise has prematurely faded.

All is wrought of fragility. A factory for killing chickens beside the remains of human bodies in a rebel-controlled field in the Donestsk Oblast of Ukraine; a place lacking sufficient morgues for so many people, such charred remains. And this is the world.

Where philosophy, heartfelt haikus and inextricable collapse in all sectors is everyday concluding – the data streaming over us with hurricane force as perdurable night descends, metaphysics must commence. The hypothalamus may or may not react to empower survival of the individual or its community; the most multitudinous of all vertebrates, avifauna, may witness the diminution of both Grandry and Herbst corpuscles - their touch receptors; as they become anosmatic (smell-blind) in combination with the loss of their hyper-sensitive and little understood filoplumes, which could conceivably wither. Even their feathers – as in the case of those few species from New Guinea to South Carolina whose feathers have been shown to exhibit varying degrees of toxicitiy – could increase as a response to the widespread megafaunal desperation and micronutrient demise and soil turn-over rate de-accelerations accompanying invertebrate extinctions.

"African Epitaph"

Born in Joal, French West Africa in 1906, Léopold Sédar Senghor would go on in life to study in France, become a Professor there, a prolific poet, and then the first President of the newly created Republic of Senegal. He decried the invasion of Italian troops into Ethiopia in 1936 with his poem "A l'appel de la race de Saba" and in 1948 Jean-Paul Sartre wrote a stirring introduction entitled "Black Orpheus" to Senghor's

Anthologie de la nouvelle poésie nègre et malgache. Like other great African nature writers, such as our late personal friend, the Yoruba/Nigerian Amos Tutuola, Senghor's unique combination of geopolitical weariness and all the evocations of a primordial continent, came together in a poem he wrote to be used for his epitaph. UNESCO published it as a Peace Poem, placing it just after Yeats and before Neruda:

"When I'm dead, my friends, place me below Shadowy Joal,
On the hill, by the bank of the Mamanguedy, near the ear of Serpents' Sanctuary.
But place me between the Lion and ancestral Tening-Ndyae.
When I'm dead, my friends, place me beneath Portuguese Joal.
Of stones from the Fort build my tomb, and cannons will keep quiet.
Two oleanders -- white and pink -- will perfume the Signare." [*42]

And why "Orpheus" one asks? From Sartre, in 1948? It was 11 years before the film "Orfeu Negro" was made in Brazil by French director Marcel Camus from the Vinicius de Moraes adaptation of the Greek mythology. President Obama has declared it was his mother's favorite film. In Obama's autobiography, he describes watching the film, age 16, and noting in his mother's sentiments the following: "The storyline was simple: the myth of the ill-fated lovers Orpheus and Eurydice set in the favelas of Rio during carnival, in Technicolor splendour, (set) against scenic green hills, the black and brown Brazilians sang and danced and strummed guitars like carefree birds in colourful plumage." [**Dreams From My Father**, *43]

Sartre could not have inflected a more pervasive, deeply ethological myth than that of Orpheus, who plays music for the animals, and did so in more Western allegories, works of art and literature than any other figure. There could be no more fitting elegy to a great African mystic who would survive unto the 21st century, carrying with him the spirit of all the ancestors who have ever co-habited the mystery of the human presence.

As for Sartre himself, he lies in the Cemetery at Montparnasse, 3 Boulevard Edgar Quinet, Paris, a few blocks from the bustling, unsettling train station. He is there today, beneath his old friend Simone De Beauvoir, in a marmoreal tomb, often covered in the cigarette butts left by literary and philosophical pilgrims. 300,000 people have been buried at the Montparnasse Cemetery since it opened in 1824. People like Samuel Beckett and his wife, Suzanne Deschevaux-Dumesnil; Frédéric Bartholdi (1834–1904), sculptor of the Statue of Liberty, as well as the great Antarctic explorer, Jules Dumont d'Urville (1790–1842), who also, near the island of Milos in the Mediterranean, was first introduced to a statue, the Venus de Milo, found days before by a local (April 8th, 1820). There is Sartre and De Beauvoir, Mr. and Mrs. Beckett, Bartholdi and Jules d'Urville, beneath the solace of some 1,200 shade trees, mostly "sophoras, thuyas, maples, ash, lime trees and conifer" [*44]. And beyond, to all sides, and further back, the noisy congeries of everyday city life. We were staying at a hotel a block behind Sartre's tomb when, turning on the late night European news stations, we watched as the World Trade Center towers in lower Manhattan were destroyed.

The shattered dust of remaining people fleeing, the lucky ones: their refugee corticosterones (21-carbon steroid hormone) emanating as stress from the adrenal glands could go unnoticed, the thick of tragic conurbation convoluting beneath the Metropolitan Tropospheric Terrorist Winter. A world glued to the instant media flow, and confusion; a confluence of violence and densities of people that mirrors a history of the runaway train that is deep demography. Malcolm Potts, in a modest little history of family planning recently detailed the pivotal modern moments of bias that have fueled our post-demographic transition calamity of human numbers. Dr. Potts, the genius of global reproductive health studies, along with his partner Dr. Martha Campbell, based at the University of California Berkeley, is pellucid.

First, he intones, by adhering to St. Augustine's theological position on Original Sin, rather than that of Celtic theologian Pelagius, the world inherited an idea regarding women that could only reinforce a male-dominated history of Western Civilization. Then, with the establishment of the World Health Organization nearly 1,7000 years after Pelagius, Original Sin was suffuse in the arguments levied by Greece, Lebanon and Belgium against any attempt to provide contraceptive services for economically marginalized nations. This politically-sexist theocracy was cemented in the *Humana vitae* encyclical of Pope Paul VI, in 1963. "Contraception was still illegal in the state of Massachusetts," writes Potts. "In 1973 Jesse Helms, the Republican senator from Carolina, passed an Amendment to the Foreign Assistance Act prohibiting the use of funds for the performance of abortion 'as a method of family planning.'"

Potts has encapsulated, under the uncanny and reliable augury of Robert Frost's poem "The Road Not Taken" [*46] a brief history of our downfall.

All these circumlocutions account for the seemingly intractable and abstruse flummoxing of any integrity or stability within the broad strokes of modern-day urban and ecologically ancestral hearths. This blurring of crucial natural selective borders are at the core of our dilemma. Spoken of differently, it profits not man or woman, whether he or she be born a penguin, a leopard, a chicken or a tree. The questions, the answers, all those possibilities, are mirrors to which those of us occupying the 21st century had better face up.

Footnotes:

*1 See John McCannon, Long Island University, "Video Review of "The War Symphonies: Shostakovich Against Stalin," Rhombus Media/Bullfrog Films, 1997. Director, Larry Weinstein. Music by the Netherlands Radio Philharmonic and the Mariinsky Theater Orchestra; conductor

Valery Gergiev. 82-minute VHS Video. Copyright © 1999 by The Journal for MultiMedia History, Volume 2, www.albany.edu/jmmh/vol2no1/warsymphonies.html.

*2 See Metaphysics Research Lab, CSLI, Stanford University, Stanford Encyclopedia of Philosophy, http://plato.stanford. edu/entries/metaphysics/

*3 See **New Views On An Old Planet – Continental Drift And The History Of The Earth**, by Tjeerd H. van Andel, Cambridge University Press, New York, N.Y., 1985, p. 271.

*4 See A. Markov and A. Korotayev, "Hyperbolic growth of marine and continental biodiversity through the Phanerozoic and community evolution," Journal of General Biology, Volume 69, 2008. N 3, pp.175-194.

*5 See **Wilderness – Earth's Last Wild Places**, Edited by Patricio Robles Gil, Written by Russell A. Mittermeier, Cristina Goettsch Mittermeier, John Pilgrim, Gustavo Fonseca, William R. Konstant and Thomas Brooks, Foreword by Gordon Moore, CEMEX, S.A. de G.V., 2002, p.28.

*6 **A Guided Tour Of The Living Cell**, Volume One, by Christian de Duve, Illustrated by Neil O. Hardy, Scientific American Library, New York, N.Y., 1984, p.1.

*7 See **What is Life? The Physical Aspect Of The Living Cell & Mind And Matter, Based on lectures delivered under the auspices of the Dublin Institute of Advanced Studies at Trinity College**, Dublin, in February 1943, by Erwin Schrödinger, Cambridge University Press, New York, N.Y., 1967, pp.7-8.

*8 See "The Transition to Agriculture and Some of its Consequences," by L. L. Cavalli-Sforza, Commentary by John F. Eisenberg, in **How Humans Adapt –A Biocultural Odyssey**, Edited by Donald J. Ortner, Foreword by S. Dillon Ripley, Smithsonian International Symposia Series, Smithsonian Institution Press, Washington, D. C., 1983, p.107.

*9 See Ralph S. Solecki, Rose L. Sollecki and Anagnostis P. Agelarakis, **The Proto-Neolithic Cemetery in Shanidar Cave**, Texas A&M University Press, College Station, Texas, 2004; see also "The Bekhme dam project in Kurdistan Iraq: a threat to the archaeology of the upper Zabros River Valley," From International Journal of Kurdish Studies, 1/1/05, by Ralph S. Solecki.

*10 **Nausea**, by Jean-Paul Sartre, Translated from the French by Lloyd Alexander, Introduction by Hayden Carruth, A New Directions Paperbook, New York, N.Y., 1964, p.99. First published in 1938 by Librairie Gallimard, as **La Nausée**.

*11 Translated with commentary by Christopher Key Chapple, in his essay "The Yamuna: Recapturing the Connection," January 3rd, 2011, presented at "The Yamuna River: A Confluence of Waters, A Crisis of Need" conference, TERI University, New Delhi. See the conference website and The Vrindaban News.

*12 ibid., p.vii.

*13 Translated by Hazel E. Barnes, A Quokka Book, Published by Pocket Books, July 1978.

*14 H.Barnes, ibid, p.321.

*15 See E. M. Cioran, **The Temptation to Exist,** Translated from the French by Richard Howard, Introduction by Susan Sontag, Quadrangle Books, Inc., New York, N. Y., 1968, p.127. First published as **La Tentation d'Exister** by Librairie Gallimard, Paris, 1956.

*16 See N. Rowe, **The Pictorial Guide to the Living Primates**, Pogonias Press, Charlestown, Rhode Island, 1966.

*17 **The Cambridge Encyclopedia of Ornithology**, Edited by Michael Brooke and Tim Birkhead, Cambridge, 1991, p.2.

*18 http://www.orionmagazine.org/index.php/articles/article/7146.

*19 See http://boards.straightdope.com/sdmb/archive/index.php/t-210885.html

*20 See "What will survive of us is love," a beautiful essay by Nigel Saul, Country Life magazine, December 29, 2010, pp.72-73. See also www.churchmonumentssociety.org and www.mbs-brasses.co.uk.

*21 See "UN's world population estimates raise questions on sustainability," Dec 14, 2010 by Michael Krebs, See http://www.digitaljournal.com/article/301443#ixzz1BD7KmqBs).

*22 **The Metaphysic of Ethics**, Immanuel Kant, Translated by J. W. Semple, Introduction by rev. Henry Calderwood, Edinburgh: T. & T. Clark, 1869, pp.41-42.

*23 See Kant, Immanuel, **The Metaphysics of Morals**, part II (The Metaphysical Principles of the Doctrine of Virtue), paras 16 and 17; See also Immanuel Kant, **Lecture on Ethics**, L. Infield (trans.) HarperTorchbooks 1963, p. 239.

*24 Jean-Paul Sartre, **Existentialism**, Translated by Bernard Frechtman, Philosophical Library, New York, N.Y., 1947, p.59.

*25 See his poem "Epistle to the Beautiful Uranie," "Believe that the eternal wisdom of the Most High/ Has graven with his hand, in the depths of your heart, The Religion of Nature." **Voltaire –Genius Of Mockery**, by Victor Thaddeus, Brentano's Publishers, New York, N.Y., 1928, p.65.

*26 op. cit., **Existentialism**, p.76.

*27 **I And Thou**, Authorised English Translation by the Rev. Ronald Gregor Smith, B.D., Charles Scribner's Sons, New York, Eighth Impression, April 1955.

*28 ibid., p.8.

*29 London: Griffith And Farran, 1869, p.106.

*30 "Saint Francis And The Birds," in **Death Of A Naturalist**, Oxford University Press, New York, N.Y., 1966, p.53.

*31 McClure, C.J.W. et al. 2013. An experimental investigation into the effects of traffic noise on distributions of birds: avoiding the phantom road. Proceedings of the Royal Society B doi: 10.1098/rspb.2013.2290, in "Birds stay away from 'phantom road'", November 7, 2013, Conservation This Week, November 13, 2013.

*32 http://www.dailymail.co.uk/news/article-2502652/Former-stray-Rupee-dog-climb-Mount-Everest.html

*33 http://protectedareas.databasin.org/pages/about-protected-areas-center

*34 www.orangutan.org/how-to-help/legacy-forest.

*35 Cook O. F. (1906). "Factors of species-formation". Science 23 (587): 506–507. doi:10.1126/science.23.587.506. PMID 17789700.

*36 "Invertebrate numbers nearly halve as human population doubles," See: http://phys.org/news/2014-07-invertebrate-halve-human-population.html

*37 "An Alternative Species Taxonomy Of The Birds Of Mexico," by Adolfo G. Navarro-Sigüenza and A. Townsend Peterson, Biota Neotropica v4 (n2) http://www.biotaneotropica.org.br/v4n2/pt/abstract?taxonomic-review+BN03504022004

*38 "Systematics and bird conservation policies: the importance of species limits," by Octavio R. Rojas-Soto, Adolfo G. Navarro-Sigüenza and Alejandro Espinosa De Los Monteros, Bird Conservation International (2010) 20:176–185. a BirdLife International, 2009 doi:10.1017/S0959270909990268

*39 ibid, from Summary.

*40 See "Shinto Deities in Japan," http://www.onmarkproductions.com/html/shinto-deities.html

*41 See **The Religion of the Nabataeans – A Conspectus**, by John F. Healey, K. Brill NV, Leiden, The Netherlands, 2001, p.143. For an intriguing list of "protection spells," and their associated deities, including Celtic, Norse, Olympian and other Egyptian, see http://www.pagannews.com/cgi-bin/spells2.pl?Protection

*42 One of UNESCO's "Poems For Peace," published on the UNESCO website between Yeats and Neruda: http://www.unesco.org/new/index.php?id=91995

*43 **Dreams From My Father**, by Barack Obama, Times Books, 1995. See Peter Bradshaw's Guardian essay, http://www.theguardian.com/film/filmblog/2009/feb/02/barack-obama-black-orpheus

*44 http://www.paris.fr/english/heritage-and-sights/cemeteries/montparnasse-cemetery/rub_8222_stand_34190_port_19019

*45 See "The Roads Not Taken," by Malcolm Potts, pp.25-29, in **Facing the Population Challenge – Wisdom from the Elders**, Edited by Marilyn Hempel, Blue Planet United, Redlands, CA, 2014.

*46 See Robert Frost's poem, "The Road Not Taken," at http://www.poetryfoundation.org/poem/173536

2

BEYOND THE ONE AND THE MANY

"Annuit Coeptis"

The United States is a nation that – among so many attributes and works-in-progress – has seriously foundered with respect to its earliest veneration of animals. If there is one absolute shift that affects all parties, it is this abnegation of America's original instincts, her earliest convictions about the wilderness. By that constituency, we refer to the Europeans occupying North America, not Native Americans. Among the colonizers the notion of gun control, eating control, modest consumption in general was slow to catch on, beginning with individuals like Benjamin Franklin and culminating in the Transcendentalist Movement whose chief adherent, Thoreau, embodied what would become the complex of American environmentalism, with all of its continuing contradictions. Pythagoras, Leonardo da Vinci, Gandhi, Thoreau, Bernard Shaw, Tolstoy… they, too, lacked the consistency of a Jain monk; which is to say, human nature is not an absolute.

We were originally a nation that loved nature unabashedly, goes the historic mythology. Consider the U.S. Congressional seal, "E Pluribus Unum" ("Out of many, one"), anointed July 4th, 1776. Its driving force is at once symbolic of what many today would like to think of as a national predilection towards a non-exclusive kingdom, a civil union, a biophilia nation: that inchoate affection, deeply moved, for all species that co-habit this world with we humans. Such romance is, sadly, restless fiction. The seal incorporates the German/Austrian two-headed eagle, the French fleur-de-lis, the Dutch lion, Scottish thistle and English rose, in addition to a man in armor, a woman holding the trident of liberty and thirteen small shields representing the original thirteen states of America. Benjamin Franklin – a vegetarian - subsequently adopted as an idea for the reverse of the seal a passage and image from the Old Testament suggesting that the Divine Presence helped the Israelites escape the Egyptians, the Red Sea overwhelming Pharaoh and his army. The motto read, "Rebellion To Tyrants Is Obedience To God." The history of Christianity is rife with Christ wearing armor; good Christians bearing arms. Thomas Jefferson, who described his own diet as a temperate one, "little animal food," his "principle diet" comprised primarily of vegetables (letter of March 21, 1819 to Dr. Vine Utley, published by the Library of America, 1984) took a likeness of this image (the original has never been found) for his own personal stationary. The actual reverse side of the so-called "Great Seal" became "God has favored our undertakings," Annuit Coeptis. This motto, which is seen above the famous all-seeing Eye, was taken at the suggestion of Charles Thomson (1729-1824) on June 20, 1782.

Thomson was for fifteen years the Secretary of the Continental Congress, the man John Adams called "the life and cause of liberty". The Continental Congress agreed upon it and to this day the United States adheres in principle to its expressly political views, informed partly by Thomson's meeting as a young diplomat of 29 with the chiefs of the Six (Indian) Nations at Easton, Pennsylvania in November, 1758. [*1]

It is indeed a curiosity of history that Thomson was philosophically moved by that fateful gathering, in as much as it was more about contest than concurrence, the French and British killing each other by a policy of divide and conquer, American colonists and native Indians being the victims. The leaders of the Mohawk nation were in attendance that November day and their leader, Nickus Karaghtadie is said to have lashed out at the policies of various war parties. Present were Bear and Wolf clan chiefs, wives, sons and daughters; a motley assemblage of Delaware and Seneca loyalists to the British. But all of this history does little to resurrect or clarify the origins of "the one and the many" ("America acts as one or not at all," said President Obama in his January, 2011 State of the Union Address) or of that league of colonial and Indian troops who were resolved to see their Six Nations rid the vast region of the American Great Lakes of French interlopers. [*2] Were these motives political, economic, religious, emotional, a combination of each? It is hard to imagine that some God was indeed favoring all of the competing causes and campaigns of destruction that characterized the chaos of European imperial interests clashing with tens-of-thousands of years of indigenous cultural heritage across North America.

"A Life of Aristotle"

One thing is certain: Fur trading, the slaughter of native species, had much to do with the ebb and flow of sympathies and the enormous financial gains to be had by the five principal constituencies in this wilderness. Those game shifters were French, British, Indian, American colonists, and naïve homesteaders all caught up in the scuttles, intrigues and violence of overreaching empires.

When Jerusalem fell to the Romans in AD 70, the early Church sent its own minions throughout Asia Minor in a similar zealous outreach of conquest. The Era of Retreat, that outmigration of monks from Egypt into the deserts of the Thebaid and

Sinai, laid the spiritual groundwork for political domination by Church functionaries who were, in fact, rebels in their own right. All the early imagery that persists with respect to the "Rebellion To Tyrants" motto is replete with swords, drowning horses, burning bushes, tidal waves of the Red Sea and a religious underpinning that, with not too much hindsight recommends nothing about the human spirit. Rather, our history appears steeped in self-destruction. The One succeeds only by conquering the Many.

This notion guided Hannah Arendt's forceful two-volume **The Life Of The Mind** where she cited Plato from Statesman, "Every one of us is like a man who see things in a dream and thinks that he knows them perfectly and then wakes up to find that he knows nothing." [*3] Arendt was particularly acute in her description of Aristotle believing "homelessness" to be "among the great advantages of the philosopher's way of life" as conveyed in his fragmentary **Protreptikos**, or "exhortation" that sought some kind of utopia at his (by then deceased mother's) home in Chalcis where he spent his final year of life. The irony, geographically speaking, of Chalcis, is that it is not among the famed Saronic Gulf Islands, Sporades, Cyclades or Dodekanese, but the main town of the island Euboea on the straight of Evripos, on the mainland, an illusion of escape from the matters of the world and named Negroponte, or "black bridge" historically.

Writes Arendt, "Philosophers love this 'nowhere' as though it were a country (philochörein) and they desire to let all other activities go for the sake of scholazein (doing nothing, as we would say) because of the sweetness, inherent in thinking or philosophizing itself." [*4] Thoreau, Keats, Rousseau, Aristotle: each thought of their minds as a kind of monastery. Thoreau recognized the impossibility of ever being lonely – with all the stars overhead. Homelessness may be a badge of honor for asceticism (indeed, it is fundamental to most ascetic traditions), but currently, at least, in Hungary it is illegal, while in nearly every state in America, there are a barrage of statutes criminalizing homelessness. Aristotle (384 BC – 322 BC) died one year after moving home to Chalcis (an ancient town first mentioned in Homer's Iliad) at the age of 62. Not one living person – then, or now – can comprehend what this man achieved, or whether – in his mind – he had found his resting place in anonymity. We have no points of reference to even begin to assess the questions he asked of the world; the answers he sought; the context he believed; or the richness of the Cosmos he had so taken to heart. It is impossible to know him; only to marvel at his genius; at his all-encompassing mind with its Four Causes, Scientific Method, Five Elements as well as the conceptualization of species versus individuals. That is a notion that would not again surface in the history of science until the early botanical manuscripts of the Swedish taxonomist, Carl Linnaeus (1707-1778) who managed to attribute sexuality, of a sort, by the late 1730s to rocks; and mistook a rhino for a rodent, but still accomplished what no formalist had ever achieved, a coherent formula for naming the Creation, where no such formula was needed; no coherence missed. Barely one second, in biological time, separated Aristotle from Linnaeus.

As for the metaphysical disposition throughout all of Aristotle's prolific life, its primary parts he likened to "being in the highest degree of abstraction" and it would dissolve into the legacy of Aristotle, a man who some have claimed contained within himself more knowledge than has ever been possessed by any other human being, over two hundred written works, according to historian of ideas Stefan Stenudd. [*5] These were not simple novels, sci-fi, short stories, movie scripts written in a weekend. Bertrand Russell would declare that "every serious intellectual advance has had to begin with an attack on some Aristotelian doctrine." [*6]

What was Aristotle's last year like? His father, Nicomachus, a doctor and surgeon in an era when house calls were the essence of medicine, had died when Aristotle was a child. His student, Alexander the Great, had expired a year before (323 BC) and posterity does not have a clue as to what Aristotle taught the young and impetuous soldier of fortune. It is known that Aristotle had a stomach problem, that the family home to which he retired to escape the anti-Macedonian hostilities circulating throughout Athens belonged to his mother, Phaestis who it is believed had died years before. The Aristotelian scholar Jonathan Barnes speculated on the great philosopher's personality and attributed admirability to him, as opposed to amiability. Aristotle's last will and testament would refer to what had been a happy family life; to his desire that his children and servants be well looked after. But beyond that, we can invoke the historical picture of his sarcasms, and brilliant persuasiveness. Nothing more.

Imagine his late night discussions with friends during that last year. There were ships at the port of Avlis where goods were loaded and unloaded from all over the Mediterranean; purple pigments were the rage, in combination with a proliferation of pottery. Flower pots. Flowers everywhere. A happy town – as if something from Garzoni Gardens in Tuscany, with its Collodi Butterfly House , the living world of Pinocchio - pictured as a veritable paradise in a late 16th century Venetian map of the Renaissance when the town was known as Chalkis.

And what did he think about? What ecological questions would such a man have asked, aside from the obvious, historically-documented queries of morphology, connections of limb, eye, vein, joint, mobility in general, breath, appendages, wings, muscles?

But, on the broader side of a forest or Mount Olympus, Aristotle had already stated that there was no such thing as "infinity" [*7] During that final year of his life it is doubtful that this behemoth of knowledge was too concerned about anything other than his poor digestion, possibly Irritable Bowel Syndrome, or stomach cancer or, as Georgios Anagnostopoulos suggests, "a chronic intestinal condition." [*8]

Did he sojourn along the walls of the port, watching a sunset, dining on goat cheese, Spanish smoked fish that came in with the ships, caviar from the Black Sea? Olives were abundant, and wine that had been "strained through cloth". There is debate amongst scholars as to whether he was a vegetarian, or not. But because Aristotle was instrumental in conveying aspects of Pythagorean theory, he may well have been. His belief that there was a certain natural servitude exhibited by plants and animals in the company of men profoundly contradicts other Aristotelian statements to the effect that "we should approach the inquiry about each animal without aversion, knowing that in all of them there is something natural and beautiful". [*9]

One outstanding characteristic of his thought is clear: the world's first major biologist, suffering from an alimentary ailment that would kill him, was probably not eager to feast on beef, turtle or roast pork when aubergine, peppers, mushrooms, horta (cultivated greens), Koukkia (fava beans) Kolokythoanthoi (zucchini flowers with rice and herbs), Lachanosalata (shredded cabbage salad), honey, paximadhi (dried breads) and so many other non-animal foods were plentiful. [*10] Dietary preferences are all speculation.

Did Aristotle's intelligence save him from the pain of mortality and early death? It was 2,277 years before Leo Sternbach, in 1955, would happen upon what has been hailed as the first of the benzodiazepine group of sedatives, including most notably Librium and Valium. The Greek philosophers had doses of grapes, berries and herbs saturated with the single plant of the Foeniculum Genus, namely, fennel, along with wintergreen, and dissolved anise oils, a drink referred to as Ouzo, of course. Silphium (*Scorodosma foetida*) a flowering herbaceous perennial, also known as Silphion, as laserwort, laser and rosinweed (the Genus containing more than twenty known medicinal sub-species), was a famed prime mover of every Greek and African kitchen. Today, it is probably extinct in the wild, but in Aristotle's time it could be imported from Pentapolis, the eastern coast of Cyrenaica, Libya.

A fascinating history pertains to the alleged magical properties of Silphion –an herbal cure-all, delicious, an abortifacient, etc., all explaining its disappearance, according to the historian Pliny, making it the first known extinct plant within ancient Western tradition, and also the elegant symbol (taken from so much ancient Greek and Roman coinage) that is the logo for the Medicinal Plant Specialist Group within the IUCN (International Union for the Conservation of Nature). There is every reason to believe that the last supper consumed by the greatest biologist who ever lived, Aristotle, included two crucial medicinal plants, Silphion and opos: the former had not yet been over-exploited, the latter –over the counter- would be prescribed for every pain in man, woman and child up until the early 1920s. Hence, the Greek phrase, Opos Agapate Όπως Αγαπάτε – "as you like it."

The Greeks have given us the very name opos from which the Middle English word of opium is derived. One finds in the Archaeological Museum of Heraklion, capital of Crete, a sculptured goddess of poppies dating to ca. 1300-1250 BC and we know that the Greeks had access to the seed capsules of the *Papaver somniferum*, or poppy plant. Aristotle would have certainly known of this mekon, or opium poppy, as the Greeks named it. Morphine itself, not surprisingly, is also of Greek origins, Morpheus, the Greek God of Sleep. Even codeine is a Greek word, from Kodeion, "poppy head". And, as scientists Askitopoulou, Ramoutsaki and Konsolaki point out, Aristotle himself discussed anesthesia in his book **Memory and Recollection**. [*11] Anagnostopoulos cites Düring who suggested that "Aristotle left Athens in the middle of a political turmoil and died the same year, a lonely man. He had few real friends and numerous enemies." [*12] And citing Jaeger, he states, "There is something affecting in the spectacle of the exile putting his affairs in order. He is constantly calling to mind his home in Stageira and the lonely house of his parents far away…Between the lines of the sober dispositions… we read a strange language…It is the warm tone of true humanity, and at the same time of an almost terrifying gulf between him and the persons by whom he was surrounded. These words were written by a lonely man." [*13]

In his remarkable discussion of Post-Darwinian arguments regarding natural selection, Theodosius Dobzhansky pointed out that "a really solitary animal is a rare phenomenon" and that "the fittest may also be the gentlest." [*14] Aristotle's loneliness, amid a life of unprecedented complexity, must leave us pondering the imponderable. His loneliness now, so many centuries later, becomes Aristotelian loneliness, but in that, the great thinker is not alone. The history of the Eclogues genre, commencing with Virgil, has a history of solitude with its distinct companionship. Consider Samuel Palmer's haunting image of the shepherd with his flock in his 1883 rendition of Virgil. Much earlier still, the four most sympathetic Persian eclogues by 17-year old William Collins of London. The Oxford undergrad had, in an elegant heartbeat, stirred the origins of romantic poetry in England, liberating previous efforts from the "inhibiting rationality and omnipresent social tone of much Augustan poetry" as 20th century Oxford literary scholar Roger Lonsdale (**The Poems of Gray, Collins and Goldsmith**, 1969) had characterized the young poet's efforts. Collins would, later in his lifetime, describe his first published work as "Irish" not Persian pastorals.

"Shamans in Exile"

In contemplating what Spanish philosopher Miguel de Unamuno had titled **Del Sentimiento Trágico de la Vida** (Macmillan Publishers, London, 1921) anthropologist Paul Radin described the remarkable differentiation in American Winnebago Indian thought between a myth and a tale. In a tale, relates Radin, though the hero was described as "being slowly boiled in a cauldron" the interpreter made it clear that Radin should not fear for the hero's life.

"Don't worry, he isn't dead. They never die," says the interpreter. [*15]

Myths, on the other hand, take in a greater span of life's adventures and realities. Tales are now, is the implication. For Unamuno, who would be banished to Fuerteventura (Canary Islands) by the dictator General Miguel Primo de Riviera in 1924, he spoke intimately of "men burdened with wisdom rather than with knowledge" and was convinced it would be impossible to ever find "truth in life" during one's own stay on this Earth, an eternal ambiguity confronted by Lewis Mumford in his great book, **In the Name of Sanity** [*16] which proffers reflection on the fact that – from where he sat – there had transpired just in the previous three decades (what would count as a Winnebago "tale" all too real) "warfare, barbarism, and extermination." And he continued, "...and even if we were not now possibly on the brink of racial suicide, thanks to the fact that through the knowledge of biotic and atomic energies we now command, we in the United States alone, to look no further afield, hold powers of life and death over mankind: separated from their use by only the thinnest walls of habit, taboo, intelligence, and moral enlightenment." [*17]

The outrage of the philosopher, the anthropologist, the author; the sullen disengagement of the great biologist, are all forms of activism. Unamuno denounced Spain's Nationalist general Millán Astray in 1936, dethroning the military's favorite La Legion slogan "Long live death!" Because of the prominence of his denunciation, one historian – Antony Beevor – called it a "remarkable act of moral courage" in as much as Unamuno could easily have been killed by his convictions. [*18] And Aristotle was exiled from Athens for his own beliefs.

In the case of Thor Heyerdahl, his public letter to Kurt Waldheim in 1978 on the occasion of his deliberately burning down his boat, the Tigris in the port of the Republic of Djibouti as an act of protest against the wars surrounding the Red Sea, stands out as one of the great transformations of this moral outrage in the face of human violence. He likened the fragile reeds of his world-roaming little boat to the world itself, conjoining one fragility with another, and imploring supposedly civilized humanity to cease and desist its self-destruction, which was carrying the whole planet in a downward spiral. [*19]

Another Norwegian, Arne Naess, who lived in the highest house in his country of Norway and once protested the building of a dam by attaching with pitons his big-wall hammock to a cliff targeted at destruction, founded the deep ecology movement. Deep protection, one might add.

Mircea Eliade, in discussing "The Shamanic Séance in North America" recognized that there were two types of problems confronted on any sickbed. First, that of the "pathogenic object" intruding upon the patient; the second, what he termed "the patient's fugitive soul" or "soul loss". Only a shaman could cope with soul loss, Eliade observed, and this treatment not infrequently involved the "summoning up" of helpful spirits. [*20] In this latter case, the activism of a Heyerdhal, a Unamuno, or any of those countless heroes who have risked their lives for a greater good, risked something of themselves, calls upon that strange bedlam of uncertainty that plagues all philosophy, anthropology, history with the same constellation of uncertainties that enshroud so many of the greatest human intellects of all time, from Aristotle to Rembrandt to Handel. This all-expressive cloud of unknowing also beckons the seemingly impossible truism of apprehending "The Other" in the case of other species, other entities, other individuals with whom we might share not merely language, but a much deeper level of artistic and emotional communion that has been taken in so many translations, "…the peace of God, which surpasses all understanding…" We (the Authors) give full presumption to the belief in the notion that each and every individual among all the other species, holds their own intellectual, spiritual and psychological power. It is equal to, if it does not exceed by unfathomed gauntlets and particulars, our own normative ethics and belief in the possibilities of humanity.

This "peace of God…." in the form of The Other, cannot be simply screened through the equivalent of a Chamberland-Pasteur filter, to determine the range of virions (virus particles) implicit in the evolution of the five presumed kingdoms of life (monera, protists, fungi, plants and animals); nor can we in good faith or intellectual integrity lend even the slightest weight to an emphasis on ourselves, amid so many Others; or differentiate ourselves with favorable sightings, origins, compass or morale when the harmony of nature has given us everything we call "great" about life, and our interpretations of, and feelings about, being alive. So obvious are the callings of biophilia as to uproot every other scientific paradigm.

"A Moonlit Garden in Verona"

All our finest poets and philosophers approach by some route of personal joy and hardship this appeal, that we should somehow stand "tiptoe on the misty mountain tops…. be gone and live, or stay and die," as Shakespeare wrote of that moonlit garden in Verona, and a lover from within. Ovid claimed some legendary Sulmo in the Abruzzo as his birthplace; Virgil, a ditch in a village that is no more, but once was called Cipada, on the outskirts of Mantua. Juvenal, in the end, sought Aquinum, his Aquinum, in Lazio, near Cassino, a short hike to the castle of Roccasecca, in which St. Thomas Aquinas would be born. Somewhere the "fairest thing on earth" exists," said Virgil. [*21]

Horace's villa looked out upon Tivoli. Legendary emperors and kings built for themselves a "golden house" within a cliff, as ephemeral as their dominions, one by one; on fated civilization after another; though the springs of Clitumnus still flow. Every home has its legacy, its champion, a caretaker without a name whose only vision has been to manage wisely its vines, let every bud have her chance. Virgil's **Eclogues** (or **Bucolics**, 44-38 BCE) address many of the caretaker's friends: the Holm Oak, Moon Trefoil, the Way-Faring Tree, Wild Tyme and Great Bind Weed; Purple Martagon and White Lilies; Stode gilliflower; White Narcissus; Common Dill, Flax-Leaved

Daphne and the gorgeous *Calendula Arvensis* (Field Marigold); Common Myrtle and Smooth Acanthus; Pale Hairy Crowfoot (the Latin sounding so much sweeter, *Ramunculus Hirsutus*); and Dwarf Edler, the magestic *Sambucus ebulus.* [*22]

These are the real homes, the fairest things, perennial fonts and source of all nostalgia. It never gets any better, no matter what they tell you. Keats did remind us of our good luck at this very moment; at the beginning of his "Ode on a Grecian Urn," writing, "...In Tempe or the dales of Arcady? What men or gods are these? What maidens loth? What mad pursuit? What struggle to escape? What pipes and timbrels? What wild ecstasy?" [*23] Or differently addressed, "You know, my Friends, how long since in my House/For a new Marriage I did make Carouse; Divorced old barren Reason from my Bed, And took the Daughter of the Vine to Spouse." [*24] No page, no date, a modern edition of a vastly re-published polymath, born 1048 AD, died 1131 AD - Omar Khayyám - made particularly famous in the West by Fitzgerald's (1809-1883) stunning adaptations of the meager quatrains or rubaiyaas. The world philosopher, poet, mathematician (the Khayyám-Saccheri quadrilateral) and astronomer (he created a map of the stars – or topography, now lost) was prolific by all accounts. His tomb is well known, in Neishapur, Iran, where he was born, his full name (offspring of tent makers) being Ghiyath al-Din Abu'l-Fath Umar ibn Ibrahim Al-Nishapuri al-Khayyami (Persian: خىام ابراهىم بن عمر الفتح ابو الدىن غىاث).

Like so many before and afterwards, Khayyám explores the one and the many; how he arrived in the form of water, and left like the wind; wrote with living fingers, not knowing why or how, then moved on – noting that neither sobriety, spirituality, not even laughter can stop the unceasing brutality that is our one life. Tears and words compete for emptiness. His vision was startling, but not unexpected. It lends itself to every melancholy, rationalism, equation (circumference, diameter, intersection, hyperbole, calendar, astronomical configuration, the birth and death of stars....) In short, a Persian version of Aristotle.

But as with the gardener in Virgil, Khayyámi verse owes its existence to the pre-Socratics, particularly the few remnants of Parmenides: 146 "complete" hexameter lines in addition to nine fragments and six Latin translations of authenticated poems by the Eleatic (Southern Italian) naturalist, who wrote in the so-called "Homeric-Hesiodic hexameter" which did not distinguish between logic and myth, religion and fact, parataxis, poetry that is enlivened by sequential subordination. [*25] It is claimed for this poet's epic "On Nature" (again, as encapsulated in these spare 161 pieces in sum) his most enduring insight concerns a remarkably unremarkable point of logic or rhetoric, namely, that what is, is, and cannot be anything but itself. Mourelatos' detective story hinges upon the psycho-linguistics of all the signs exhibited in the brief, flamboyant journey posed by Parmenides, ingeniously translated into English by Arthur Fairbanks. [*26]

"The Rites of Parmenides"

In this mythic journey, horses bear the character of Parmenides; he was born to ride that chariot, driven by the "maiden daughters of Helios" towards that "gate between the ways of day and night" which open according to "retributive Justice". [*27] In the manner of ancient Jain tradition, Parmenides accepts no beginning or end, confirming the eternal return along a "single path" and obsessed with the fact that there can be no declaration, thought, or truth if it dares to express the notion of "not-being". [*28] Being may "suffer destruction" and whole generations can be extinguished according to this premise. But, at the same time, Parmenides is fraught with undecided ambiguities between being and non-being. Hence, his statement that "there is the aethereal flame of fire, fine, rarefied, everywhere identical with itself and not identical with its opposite.....(that) all things are called light and darkness..."[*29] And that the underlying assertiveness of the universe commends, first and foremost, "love," then "the wandering deeds of the round-eyed moon and its nature" and finally the fact "she is always gazing earnestly toward the rays of the sun."[*30]

Much of this echoes seven hundred years later in the writings of St. Basil (born in Cappadocia around 329 AD) as well as many of the monks at Mount Athos who would speak of the vulnerability of the soul, imperceptible incitements, and – the most important of all – that "first step" towards "retired habitation". [*31]

The echoes are rich, and their beginnings – from Thales to Pythagoras (allegedly the first animal liberationist, going out to local markets and purchasing animals so as to free them) [*32] – bespeak of a paradigm fully ingrained in the human psyche. It did not start with Thales, to be sure. André Leroi-Gourhan (1911-1986), one of the premier paleoanthropologists of the 20th century, best known for his interpretive efforts within the world of cave painting and the concept of genetic memories, as well as the "liberation of memory" provided ample proof throughout his copious work of a clear veneration of nature enshrined within such sacred sites as Lascaux and, much later, the newly designated UNESCO World Heritage Chauvet. But the nimbus, 2, 3 or 4 color enigmas, the nature of mixing pigment, Mayan glyph translations, the zoosemiotics of so many hybrid stick figures, portals, apparent templates of proximal landscapes (as we have viewed in the hundreds of 3-color images just south of Burgos in northeastern Mexico) all speak to a protective, custodial impulse that is as pragmatic as it is imaginative.

The Burgos phenomena is believed to be a mere 120 generations removed from today, comprising two, possibly three nomadic Native North American tribes whose range of activities cannot simply be defined by the pictorial record of other species, tools or landscapes. Today's ecosystems in the state of Tamaulipas are probably unchanged in the subsequent 5,000 years; some of the most dense and diverse butterfly populations on Earth; dry mesic oak forests intersecting northerly neo-tropical jungle; precipitation coming from the Gulf of Mexico, and no hint – for 4,500 years – of incoming Spanish colonials intent upon reshaping the geopolitics of the region, or burning down a sacred "House of Birds" in Montezuma's Aztec capital of Tenochtitlan. Hence, the suggestion of a true paradise for peoples that appreciated everything around them. Were they intent upon protecting their surroundings, scant possessions, one another? Human nature being what we now believe it to be would indicate some semblance of secrecy for purposes of discretion; weapons figured in some of the rock art tells whatever story we are disposed to read of it. There is no outright killing, as seen in certain Neolithic cave art from present Namibia and South Africa; no spears flying, heads hanging, or pursuit of wild animals.

Burgos presents, instead, a period of relative tranquility that stands out in the human record on Earth and forces us to readjust our sights before proposing general rules of a brutal nature, as has been the Darwinian biological ethos, taken to an ethical key-point by Herbert Spencer (1820-1903) who, upon reading Darwin's **On the Origin of Species** would invent the phrase "survival of the fittest" in his book **Principles of Biology** (1864). Because Spencer applied natural selection not only to biology, but to ethics, government, the human race both at the stage of individuals, and in aggregate; his formidable presence in the history of science and sociology, political philosophy, economics, psychology, and natural history has always come as something of a shock to newcomers to his work. In his 1857 essay "Progress: Its Law and Cause", published in the Westminster Review, Spencer indeed – like Alfred Wallace writing under the influence of Malaria in Java – described a conceptual world in which evolutionary forces played a dominant role. This, two years prior to Darwin's overnight celebrity. But, as Spencer – forever maimed, and by 1930 all but forgotten, as the "social Darwinian" – had indicated, there was an inevitable convergence of individual and of species. Moreover, according to Spencer, natural selection must necessarily inform our morals. Such was Spencer's sixty-year career of actively articulating an approach to an enlightened and ethical utilitarianism that could enshrine all future human perfectibility.

But he was also personally frail, an insomniac, which cannot possibly help, and prone to nervous breakdown. In short, his waning physical health, and the early death of his many siblings seems to have cast a shadow over Spencer's otherwise deep appreciation of natural wonders and their utilitarian tonic. Yet, he was a pessimist near the end, seen wandering the streets of London in fits of deep despair. Famous, but unread; influential to an ever diminishing group of cohorts. He seems to have perceived little future for humankind.

Prior to the onrush of evolutionary theory and the worrying that subsequently beset all the sciences and social sciences concerned with the Commonweal and its relations with those Others, it is clear that a working model of humanity prior to the invasion of science can be surmised from the innocence displayed in all its fidelity at the eleven known petroglyph sites surrounding Burgos.

Across Western and Eastern Europe, as well as the period of *Homo erectus* in Southeast Asia and Micronesia, because so many of the cave labyrinths were of such tortuous geologic formation, it is likely that the sacredness – upon whose origins, teleology, guises, technology and nuances hinges a vast speculative literature – hovered around not just the secrecy of the sites, and of the petroglyphs, but the nature of aesthetic and emotional preservation. The painters and shamans involved in the expression of animist relations and essence seem, by so many means of articulation, color, psychological address and reciprocity, to have endowed the life force with a privacy akin to worship; a reverence that was personal, the heart of protectionist instincts.

By the era of Parmenides, said to have been born sometime between 540 and 515 BCE, we see a synthesis of these instincts diffused through countless schools of thought. That there is one, and there are many. Do they meet? Do they speak, understand one another? Are they sentient entities, or objects of the abstract? And in whose estimate? Every opposite can be posited for their juxtapositions in "On Nature." Mourelatos refers to some of them: "perceptible appearance vs. inner being; duality vs. unity; otherness vs. sameness; growth vs. no growth, and so forth. [*33]

None of these numeric references are mathematical in the sense of probability theory, however. There is no cumulative distribution; no Fourier transform; neither Bayesian statistic, quadratic function, exponential nor Gaussian analysis. And if these hypocatastases bear no resemblence to primeval pre-Socratic pantheism, or jump off the page, then think, instead, of the prosody that science, in her least methodological aspirations, invites: consideration of a rhapsodic medley that is the world Thales knew; the Homeric seas, cliffs, "wine dark waters," and so forth.

In speaking to life, love letters to ourselves, more important, to our friends, no standard unit of distribution is relevant; there is no equivalent geodesic survey of the Kingdom of Hanover (Gauss' 1818 heliotropic method of measuring distances using a mirror and the reflection of sunlight) that can summon forth a probability factor that is both biological and non-Euclidean, as proved by Eugenio Beltrami in 1868. The pseudosphere, as he called it, made for multidimensional understanding and laid the groundwork for circular reasoning; for such inventive fictions as **Alice's Adventures In Wonderland** (a rabbit with a waist-coat and a pocket watch; a rabbit hole; a first question – is this the New Zealand or Australia of 1878? And **Through the Looking-Glass**; the Jubjub bird and frumious Bandersnatch; the slithy toves and Humpty Dumpty; the garden of live flowers;) and Norton Juster's **The Phantom Tollbooth**, nearly a century later. Not to mention the whole interdependent predilection of symbiosis, and the earliest concepts of the evolution of eukaryotes.

Pythagoras was a contemporary of Parmenides and a long-acknowledged vegetarian, perhaps – with Shelley, Einstein and Leonardo da Vinci- the most famous of them for reasons that soar to the quintessence of Greek dialectical thinking and its subsequent impact on religion, philosophy and all subsequent military history in the West. There is a tremendous whimsy (in retrospect) accompanying Pythagoras' school of thought. Philolaos (470 – c.385 BCE) was convinced, as Fairbanks points out, "that the earthy appearance of the moon is due to its being inhabited by animals and by plants, like those on our earth, only greater and more beautiful; for the animals on it are fifteen times as powerful, not having any sort of excrement, and their day is fifteen times as long as ours..." [*34] From such scientific/mathematical deductions to the voyage extraordinaire genre of the Renaissance, including the entire Utopian literature that fortunately persists to this day, we see the same desire for a beautiful world, an exquisite natural world marshaling its myriad manifestations in every school of thought. History records a fixation with paradise as ingrained as love and food.

The earliest Asiatic aesthetics were rife with shan/shui, or mountain water paradise imagery. Much of contemporary art only reiterates the elements of this obsessive want. In Western religious and artistic contemplation – from the Anglo-Saxon Bede (673-736), to the 14th century French **Nouvelles de la terre de Prestre Jehan**, to a wave of interpretations of how Adam and Eve spent their time, most notably in the work of the Syriac Bishop Moses Bar Cephas (813-903) whose treatise **Liber de paradiso** was reissued by the famed Christophe Plantin house of Antwerp in 1569, in the same vein as works by John Salkeld -**A Treatise of Paradise,** 1617, and Marmaduke Carver, **A Discourse Of The Terrestrial Paradise Aiming at a more probable DISCOVERY OF [sic] The true SITUATION [sic] of that happy place of our First Parents Habitation**, (London, 1666).

The co-evolution of elaborate botanical gardens, the search for Noah's Ark, moreover, the essence of Noah the man, and the increasing fascination with "specimens" representative of the Original Garden of Eden, the basis for so much artistic redaction and euphoria, translating into every variety of scientific discourse, all gave loft to this endlessly dreaming hope that humanity might somehow be enlightened and reconnected to its original Utopian circumstances. 21st century discourse has discounted, indeed –ridiculed, all things Utopian, Lamarckian, as somehow Luddite nonsense, of the Ken Kesey, Jack Kerouac era. Forgetting, of course, the massive and majestic oeuvre of such poets as William Everson (Brother Antoninus (1912-1994) except at the William Andrews Clark Memorial Library of UCLA, and the rare books room at the University of California-Santa Cruz, where so much of Everson's earth poetry are held sacred.

Post-Holocaust speculation is without mercy for fancy. There is good reason, to be sure, amongst humans. But our conservationist patience with the debate in New Zealand over three species of rat, one mouse, one Australian possum, red and white-tailed deer; in the U.S., over some 50,000 non-native species - from the magnificent Burmese python in the Everglades to Asian Long-Horned Beetles throughout the Northeast, even the playful, Caravaggio-wild "feral" hogs whose itinerary from Eurasia to Oklahoma has somehow ignited more consternation than Oklahoma's Trail of Tears – is, in this end we now inhabit of minor import. We are whiners when it comes to the inevitability of the biological cocktail that Earth is constantly exploring. In humanity's compulsive itineraries, Gaia has found her most flamboyant and unconscious agent. Parmenides seems to have seen it all. Aristotle was less certain on some points. For example, he never imagined human civilization occupying the tropics with any staying power.

Are such speculations regarding lunar biodiversity 2400 years ago that much different than, say, the notion that "members of the dog family tend to pair off as birds do"? Or that "Coyotes and foxes, golden jackals and black-backed jackals, for instance, are all committed, long-term monogamists?" [*35] Such is the continuing fascination with earthly origins and comparisons that might somehow ease the pain of contemporary chaos, whether in Syria, the U.S. (with regard to gun violence), or Iraq.

Current "Mysterytopia" would have played well to ancient Greek audiences, with such large personalities as the Angora rabbit, Axolotl, Blobfish, Frill-necked Lizard and Pink Fairy Armadillo. [*36]

"The Myriad of Vulnerable Souls"

So many of the great natural history expeditions – from those insects painted by of Maria Sibylla Merian in 17th century Suriname, **Metamorphosis insectorum Surinamensium**, in 1705 [*37] to James Wilson's **Illustrations of Zoology, Being Representations of New, Rare, or Remarkable Subjects of the Animal Kingdom** (Edinburgh, London, 1826-31) to the wealth of botanical discoveries made by Alexander Von Humboldt throughout the world – all confirm our astounding innocence when it comes to knowing very much about the biological circumstances of our own species. We have identified fewer than 1.9 million species but suspect there to be as many as 100 million or more species across the terrestrial and aquatic domains, particularly amongst invertebrates, not to mention the bacteria and viruses inhabiting those invertebrates. To this day, the majority of "spokespersons" for zoology refer to fewer than 9 million species. It is a strange abnegation. Research in the deep waters of the Antarctic Ocean have revealed millions of new species. Similarly, at Yasuní National Park in Eastern Ecuador, more tree species in one hectare (2.47 acres) of rain forest, or 655, have been discovered than exist within all of the U.S. and Canada combined. Adding to this is the remarkable extrapolation of some 100,000 insect species in that same hectare. Multiply the 100,000 times the prospective average number of individuals per species, and we are dealing in exponential surrealism of 10 to the 10th power and beyond, all dependent on a single tree. [*38]

If conservation protection is even loosely maintained by the Ecuadorian government – and it suggests every indication of endeavoring only the same balancing act (with $14 billion worth of proved petroleum assets beneath the national park) as has devolved in every other nation (jobs versus biodiversity) – then this could be likened to a biological welfare state, such as has been beautifully dramatized in early Lao literature of the Indo-Chinese peninsula, with such protective protagonists as Thep Malai –an ascetic who traveled to paradise and back and whose tales were later transcribed in the famed Buddhist Jātaka, as well as the Laotian Kalaket full of "enchanted forests". [*39]

There is reason to be hopeful about such, otherwise peculiar, analogies when one considers that – as the venerable Irving Howe once pointed out- "Indisputably, the welfare state has eased, if not wholly liberated, the lives of millions of people." [*40]

Not unlike the Ancient Greek mathematical and ecological reveries, Jain philosophy shared this cosmological credo with respect to numbers and biology. "Parasparopagraho jīvānām," translated as "Souls render service to one another" is amplified in the writing of Umāsvāti/Umāsvāmī (2nd century BCE) when he says, "A soul can occupy space that is one innumerablth [sic] part, or more, of cosmic space. Like the light of a lamp, the soul assumes the size of the body it happens to occupy on account of the contraction and expansion of its space units." [*41] In contradistinction, Chinese and Japanese invocations of the soul occupying space are best discerned in the garden traditions which, regardless of the proverbial "Auguries of Innocence" in which William Blake wrote of "…a grain of sand, and heaven in a wild flower/Hold infinity in the palm of your hand, and eternity in an hour" (The Pickering Manuscript, 1803), Asiatic spirituality has tended towards the Persian approach to paradise, namely, placing a wall around a garden. [*42]

The Greek word, organismos, from organon, meaning a tool, did not even appear in English until 1701. Nearly every ecological concept in current usage throughout the West is of Greek origin, the three primary facets of all biodiversity stemming from Archaea, "ancient things", Bacteria - prokaryotes "before the nucleus" - and Eukaryota, "true nuclei". Natural selection has no preordained timeframe. In David Reznick's research on the relationship, for example, between commercial fishing and fish evolution, he has prompted a sounding board for empirically cataloguing the prospect of natural selection occurring ten million times faster than might be expected under previous Darwinian theory. Along with Helen Rodd and Leonard Nunney, Reznick's work has been enlightening with respect to the fresh water guppy (*Poecilia reticulate*) life histories in the Northern Range Mountains of Trinidad, where the presence of waterfalls restricts distribution of individuals and can actually trigger totally different rates of community size, population robustness, and the acceleration, or de-acceleration of predation, with all of the implications. [*43]

Every comparative phylogeny and DNA sequence yields vivid testimony to some new insight as science plows through nearly every biological premise and overturns each theory with darker or brighter hypotheses, while the actual technical origins of life entirely elude us. All we can generalize about mammalian evolution is the nature of nurturance through the method of mammary glands. But even the most elementary proclivities of compassion are called into question, as in the case of humans feeding birds during the Spring, with recent data suggesting that bird feeders may impede clutch size during breeding periods in the Northern Hemisphere. [*44]

There is a logic to this, though it fails to live up to the promise inherent in humans helping birds amidst the worst decline in avifauna in 65 million years. With this ambiguity in mind, Wittgenstein's statement that "there can never be surprises in logic" takes on a renewed power. [*45]

Indeed, Darwinian theory falls apart on ethical grounds when logic confronts the unassailable possibilities of ethical deliberation, one reason we may assume the equally unassailable John Ruskin, rich in self-contradiction, what Frederick Kirchhoff has referred to as Ruskin's "alternative taxonomy", nonetheless could argue magnificently that "the aesthetic relations of species are independent of their origin" and that "the powers of nature consult quite other ends than the mere continuance of oaks and plum trees on the earth." [*46]

"The Underestimated History"

Nurturance has a rich but underestimated history, while care giving provides insights complicated by the adult nature of its profession, and the practice itself whose externalities encompass medical insurance, hospital staff, and a host of legal, economic and political factors that are not biological, despite our recent revelations regarding the tau protein and cerebral plaques comprised largely of beta-amyloid, and the nearly 45 million humans currently suffering from some version of dementia or worse; and, arguably – if there is a human world – 150 million victims by 2050.

We keep brain-dead patients alive, or not. If an adult can get to Switzerland, pain, no pain, doctor's verdict not required, disease, or just depressed, he, she can ask to be euthanized. This is a miraculous vote of confidence for Switzerland.

We dispense pain-killers with variable generosity, from culture to culture, but are indifferent in terms of overall percentages of people, but our stupidity and selfishness towards one another is nothing like that picture which instantly emerges in any relevant setting pertaining to our slaughter of innocence, the suffering of most other animals and moribund habitat about which much of the aforementioned histories of biology have so celebrated. That, again, is the essence of Parmenides.

More to its origins, nurturance extends from the parental roles in evolution to how not only adults but children, and the child in adults, respond to babies (think of the film "Loosing Isaiah"(1995), directed by Stephen Gyllenhaal, and siblings to one another. [*47] Laura Huxley, the late widow of Aldous Huxley, through her work with young people in Los Angeles, described to us how children helping younger children in need typically discovered in themselves heretofore untapped resources of compassion. Indeed, she often quoted Aldous who had said, "One never loves enough." [*48] The same phenomenon happens amongst children who grow up with companion animals of other species in their household, a circumstance that is widespread throughout the world but comes particularly into focus at sanctuaries like Green Chimneys in Brewster New York. Their mission is that of "restoring possibilities and creating futures for children with emotional, behavioral, social and learning challenges through educational, therapeutic and outreach services, while providing care for animals and nature, thereby promoting a philosophy of dignity and worth for all living things." [*49] In filming at Green Chimneys we watched a boy from a broken home bond with a rescued pig of enormous stature (and weight). Their love affair was transformative for all concerned and underscores the magnificent truth that children raised in a collaborative environment with other species will likely discover their true essence with a greater totality than can possibly be imagined in absence of exposure to wildness: wilderness in terms of habitat, and the wildness whose mirrors embrace all creatures, including humans but more so, the vastly greater majority of others. [*50] In film director Carroll Ballard's breathtaking adaptation of biologist's Farley Mowat's 1963 amazing autobiography **Never Cry Wolf**, released 20 years later and starring Charles Martin Smith, Ballard enshrined the most compelling testimony to human contradictions with respect to nurturance ever revealed in a work of cinema.

The countless stories of bonds under the general heading of biophilia are astonishing and predictable to any naturalist, or reader of the great realm of so called childrens' literature, even if polls suggest that most children prefer the company of dogs to that of cats.

This biophilia connectivity has so quietly revolutionized the sciences and humanities as to evoke a more adequate picture, perhaps understanding of the reason many have considered outright the possibility that there is a God. That He, or She abides in all Being, past, present and future. One vivid collection of ever-growing testimonies to this evidentiary bonanza are the anecdotal visions of home videos rendered available to humans worldwide in an instant on YouTube or any other comparable digital technology. Every conceivable interspecies conviviality showcased in amazing intimacy: an Australian lad and his wild marmot friend; a dove playing with a sleeping cat; a gull fishing utilizing a piece of bread; hippos and turtles; dogs and foxes, dogs and cats; African cattle egrets and water buffalos. And, most famously, the gorilla Koko and her chosen gray male Manx kitten, All Ball. Koko's fame has also arisen because of her mastery of human sign language.

The Prophet Isaiah was instructive, but no more so than today's technological visions of the expanded purview across every continent. The history of these connections between humans and other species is so vast, encompassing so broad a literature, as to defy enumeration. A few key cornerstones of the biographical continuum should suffice: the great French tapestry, "Lady with the Unicorn," (c.1510 – also present, of course, is a conspicuous lion, dogs, rabbits, mink, ermine, etc.); the poetic landscapes of Virgil, those pictorial visions of Giovanni Bellini, Piero della Francesca, Jan Breughel the Elder, Leonardo da Vinci and Jan Vermeer; social-scientific works like Pietri Von Hartenfelss' **Elephantographia Curiosa** (1715) – the first major study of the elephant in human culture – or George Adams' **Essays on the Microscope** (1787) with its extraordinary entomological illustrations; the gorgeous art of Arthur Butler' **Foreign Finches in Captivity** (1894-1896), inciting every justified melancholy by the very realization of these captive charges, as distinguished, for example, from all those beautiful flower and tree books, from Darwin on orchids (1862), to Ellen Anne Willmott's **The Genus Rosa** (1910-1914). In contemplating the lyrical wisdom of Isaiah, one recognizes not merely the animals themselves as full Beings in every way equal, if not superior to humans, but the equally charged personalities that are the very landscapes in which the communion fueling Isaiah's most memorable passages reside: landscapes without end.

From Isaiah 11:6, wherein "The wolf will live with the lamb, the leopard will lie down with the baby goat. The calf and the yearling will be safe with the lion…" imagine the possibilities, and the physiolatry that informs those scenes, far away, beyond the horizon, and in our very backyards.

"Landscapes Without End"

Three works of deeply nurturing importance lend particular power and diversity at this juncture: **Landscape**, by Philip Gilbert Hamerton (London: Seeley & Co., 1885), Birket Foster's **Pictures of English Landscape**, India Proofs, Engraved by the Brothers Dalziel, With Pictures in Words by Tom Taylor (London: Camden Press, 1881), and **Landscape Painting – A History**, by Nils Büttner, Translated by Russell Stockman, Abbeville Press Publishers, New York and London, 2000). These three luscious tomes celebrate and rejoice in the natural world through an exploration of the arts and the imagination.

Of course "the arts and the imagination" is a thorough dud of a phrase; like some pedagogical syllabus absent vim, vigor and disposition; no meaning, really; no concept of the South Face of Dhaulagiri in Winter, on a clear night, at sunset; De Gerlache Strait, Western Peninsula, Antarctica, at sunrise; let alone a full moon, as viewed through Shipton's Arch in Xinjiang, China.

The arts, every animal's imagination, is emblematic of that same improbable but possible impulse in ourselves to disregard rules, wander without direction, and be wonderfully forever immersed in that realm that is being destroyed by the second.

How can there be such reverence in the face of a downpour of malevolence? This question must incite gout in the devotee, build up bountiful bile in any believer;

inure solemnity to outrage; give guidance, or at least a banner, to all those who would champion the salvation of that which will otherwise be pursued, chopped down, and slaughtered.

In the case of the amazing Hamerton, his unprecedented meditation on the effects of landscape on the human soul was hardly noticed when it was first published in a limited edition of roughly 525 copies. In presenting gorgeous original and reproductive etchings by Samuel Palmer, Claude, Gustave Courbet, Camille Corot, Titian, J. M. W. Turner, David Cox, Maxim Lalanne, C. F. Daubigny, Meindert Hobbema, Albert Dürer and many others, Hamerton's massive effort quite effectively captured all that tenderness of spirit that poets from Horace to Victor Hugo had iterated; mountains "shining like truth," the joy of English hop-gardens and French vineyards; the Mer de Glace and English moors.

Hamilton, who also wrote about animal rights long before whole books had been devoted to the subject, was of a fully open mind, determined, like Wordsworth who he cites, to love nature "passionately" while also managing to speak directly to the "heart of the farmer," surpassing the multitudinous bullshit that ranks amongst the most lethal and widespread levels of deceit ever institutionalized, namely, the agricultural pastime of slaughter; the notion of "husbandry" that conceals from its children the true insanity of killing in the rural sector, where it is all but reverenced.

Hamilton reconnoitered remote river banks, fell asleep under trees, laid all of the intellectual foundations for the greatest marvels in the world, captured on what, after all, are merely vulnerable pieces of paper. In his effort to follow no academic strictures or preordained rubric, he shared with posterity the simple Pure Land Buddhist truth that we can indeed become one with our passion, with globular fruit, churchyards, with "rows of magnificent oaks," even grey sky "devoid of form" amid "painful longings" for all that makes us human in the landscape. We can transform dread law, dreary docks, mechanized warehouses, the stockyards of Upton Sinclair's Chicago into a green and transcendental place.

Speaking for his fellow Englishmen, he admitted that "We, in our time, have been especially and exceptionally favoured" and owed this debt to nothing other than, for example, "a distance by Titian."

In a similar vein, the Dalziels captured etchings unlike any ever produced with respect to rural sensibility. Each image is a Church, so to speak. But the poet Tom Taylor is quick to point out that "...Nature's beauty and use and care/Are all unlike to Man's". [*51] They illustrate the "Four States Of The Stream," the nibbling on green leaves of "grazing sheep" like "those who play where churchyard grasses wave" [*52] and address "A wife's and baby's greeting" [*53] "From God's Glad and generous Earth." [*54]

Myles Birket Foster (1825-1899) exceeded like few others in visually abetting a fast-fading collective memory of England's rural sector into an historic antidote to the swift and brutal Industrial Revolution all around it. By 1863 his success with scenic watercolors, drawings and engravings had enabled him to repair to the village of Whitley, in the civil parish of the Borough of Waverley in Surrey and there have built his "Hill" house near Godalming. About that time his elegant and sentimentalized England was becoming a national fixation. Cadburys, the chocolate kings, used Foster's images on many of their boxes.

Foster painted all over Scotland, Switzerland and the Mediterranean. Two days after his death, March 27th, the last year of the "Old World," the Times of London called him "certainly the most popular water-colour artist of our time."

Foster is buried at the All Saints Church in Whitley, which dates to 1040, a few miles from the Sandhills Area of Outstanding Beauty on what is called the Green Sand Way, wherein is to be found an ancient pre-Norman Common, owned by the National Trust, and an important sanctuary for donkeys run by the RSPA.

The Brothers Dalziel who helped produce Foster's major work of engravings also published William Holman Hunt, John Everett Millais, Dante Gabriel Rossetti and James McNeill Whistler. Prior to the photographic publishing age, the Dalziel firm, which had begun in the 1830s, was considered the most accomplished and influential of the British 19th century and Foster's collaboration with them ranks as probably the artist's and the publishers' greatest achievement.

In surveying the vast and ambitious field of landscape aesthetics, Nils Büttner's scholarly overview [*55] remains the best and most thoroughly diagnosed history of the impulse, the state of the world through art; of men like Birket Foster. Büttner's mammoth undertaking might suggest an impossible passion, because landscape painting is not unlike the "broken window" syndrome of classical economics. That debate, stoked by Frédéric Bastiat in his 1850 essay "Ce qu'on voit et ce qu'on ne voit pas" **(That Which Is Seen and That Which Is Unseen)** adds to the conundrum such other obsessive ironies as the "opportunity costs of war" and the subjective theory of value, wherein value lies not in the thing itself, but in the person who imagines he/she actually owns it.

The point is that everyone knows landscapes; everyone is an expert, having been born of this Earth. The difference in the landscape aesthetic sensibility is having the moment out of time to smell the roses; to see what others have seen a million times. To feel what has not been felt before. To be so moved – whether in spiritual deification, in whimsy, reverie, poetic outpourings of the most somber, ecstatic or simply silently personal sort – by nature; transformed into a new being, a new nature that can reciprocate, or at least feel singular urgings to do so – to be part of this magnificent picture, compassionate and humbled; to help another.... All these symptoms of falling in love are at the heart of naturalism and while there are a million landscape pictures, trillions of personal biographies precisely hewn from nature and evolution, it takes an audacious resolve to attempt, however incompletely, to formulate a through-line that is cogent in all this. Pretty, divine, sullied, tortured, reclaimed, redeemed; at the crossroads of human history; an apotheosis, an apology, a severed connection; an eternal return. All these condensates of the human experience, through every filtration system known to our genetic penchants, co-create within that endless cavern of wild remnants that stir us still.

In attempting to somehow marshal the necessary coherence, thematic, linear, identifiable signature landscapes, Büttner has attended lovingly upon such moments in time as the "Landscape With River Mouth," by Cornelis Vroom (circa 1591-1661) [*56] exercising deep restraint in his characterization, but rendering clear the strength of the picture by its "actual protagonists" (the hugged trees and "their exquisitely rendered crowns"). Every painter portrayed is "longing" after that "inscrutable beyond" [*57] which, as Vincent van Gogh, in his 1890 painting of "Wheat Field With Crows," described in a letter to his brother Theo: "...in the motif of the vast wheat field he [Vincent] had tried 'to express sadness and extreme loneliness, utter loneliness...what I cannot express in words.'" [*58]

In tackling this seemingly contradictory relationship to nature, Büttner elaborates upon the art historian Wilhelm Worringer's attempt (from Worringer's book **Abstraction and Empathy**, 1908) to "distinguish between the naturally beautiful and the artistically beautiful, and set up the need for empathy and the urge toward abstraction as oppositions." [*59]

It is this correlative, enigmatically foundational relationship between our imaginative ordeals and our joy, that the landscape aesthetic emerges forever new and original. Each point on the compass commands our attention: an outcrop, a cataract, moody Scottish highland storm warnings, Himalayan abysses, Arctic white-outs, lush, Romantically neglected monasteries and monastic gardens; Burgundian picnics, Lowland estuaries, bright eyed bovines of the Barbizon, sunsets along the Hudson River, every conceivable point upon which humanity has stood, a man and a woman, beneath a tree, or upon a river bank, have fallen in love, or dreamt away their starry nights.

And in spite of all the melancholy madness implicit in this attempt to empathize amid the phenomenological evidence of bad news to date – a task well shared by artist, scientist or activist alike – landscape aesthetics ascribe elegant and universally imperative strength to the human spirit as it engages the world around it in an effort, one prays, to protect and cherish it.

This Forever embrace can have only one objective in its quiet, methodical labors, to conserve, to save what is left. If that is our legacy, let it be so; let it be the premiere understanding of ourselves, our own capacities to honor the Earth, our one and only truth, our sole honesty; the only mirror that can report back to a far-gone future and declare: human beings had a heart after all. They loved pretty things; they swooned over moody nature, were restless and impulsive before the Sublime, and waxed endlessly philosophic upon the brink of their own lives, and those of everyone around them.

"1938"

In many respects this evolutionary orientation to naturalism hits a relentless mark in the year 1938, during which four watershed events occurred. Their convergence is no coincidence but, rather, a signal fire whose confluence suggests all of the historic highs and lows of human nature. The complexities of our peculiar evolution. The worst horrors, and most fantastic daydreams our species has experienced.

First, the emergence of the tragic relationship between young Jody Baxter and his adopted fawn, who he names Flag, in Marjorie Kinnan Rawlings' breathtaking novel, **The Yearling**. There, across middling Florida, played out in heartbreaking reality, to Norman Rockwell's vivid first edition color illustrations, is the young boy and his Bambi. Both are victims of the deer and bear hunting mentality that would, in fact, drive the *Ursus americanus luteolus*, one of two distinct Mississippi black bear subspecies –the so called Louisiana Black Bear to near extinction. Half-century after Rawlings' novel, an estimated 60 or so individuals were left, in the Pearl, Pascagoula and Mississippi River washes. In the case of the Florida Key Deer, the smallest White Tailed deer in the world, there were a mere 25 known individuals as of 1955. In fact, as many as 89 known species that once existed in Florida have probably gone extinct. [*60] As of 2013, Florida's mascot, the Florida Panther (*Puma concolor coryi*) did not exceed 160 in the wild. It was a far cry from the situation when botanist William Bartram wrote of his **Travels through North and South Carolina, Georgia, East and West Florida, the Cherokee Country, etc.** between 1773 and 1777, the book published in 1791 (re-printed in London and Dublin three times during the following two years). Thousands of panthers wandered throughout much of the East, with evidence of large numbers in the southeast, where they had evolved over tens-of-thousands of years. [*61]

Indeed, there is ample evidence of prehistoric Native American artwork incorporating the panther. In 1896 the young Frank Hamilton Cushing led the Smithsonian Institution's Pepper-Hearst Expedition to a small pond on Key Marco, which became known as the "Court of the Pile Dwellers" (8CR49). There Cushing discovered a precious wooden carving of the big cat, along with more than 1,000 other artifacts, making it possibly the most important archaeological discovery on the entire Eastern Seaboard. [*62] Cushing's book, **Exploration of Ancient Key-Dweller Remains of the Gulf Coast of Florida**, first published in 1896 by the American Philosophical Society, was republished in 2000 by the University Press of Florida. Randolph J. Widmer, in his Introduction, writes, "Today it is impossible to visualize the original topography. The site has houses, condominiums, shops, and restaurants on its now-level surface, which is dissected by canals and paved streets. There is nothing visible to suggest an aboriginal site..." [*63]

Dr. Anna Lillios has written of Rawlings' beautiful descriptions of rural Cross Creek (in the Ocala National Forest) where Rawlings and her husband had moved in 1928. Consider this passage, with its reference to panthers: "And the country was beautiful, with its mysterious swamps, its palms, its great live oaks, dripping gray Spanish moss, its deer and bear and raccoons and panthers and reptiles." [*64]

Rawlings has not drawn a picture that is outside the postcard. Her portrayals of hard-working people were entirely sympathetic, acutely and sensitively observed. And she lived there – out there.

Yet, nothing beyond the exceptional brutality of North America generally speaking, from the time of its colonization by Europeans, to the same forces of expansion that would overwhelm site 8CR49, are the same forces of obfuscation and indifference that are driving the Florida panther, and so many other species to extinction.

While many have argued against the idyllic romance of indigenous peoples in North America treating wildlife any differently than those hunters who arrived en masse in places like the newly acquired Louisiana Purchase, the choice to kill, mow down habitat, dam rivers, pollute streams, foul the air, tarnish the soil, deplete aquifers, slice mountain tops and trash the marine and riparian ecosystems, remains an unequalled height of folly and destructiveness. The Native American systems of survival could not begin to approach that malignancy of colonialism that swept hundreds of tribes up in maelstroms as genocidal as that afflicting the panther.

Hunters unabashedly led the pack. Lumbermen followed. Then real estate developers (after the very early example of President George Washington himself, who is said to have owned, as of July 9, 1799, 52,200 acres in several states). Florida as a whole was symptomatically poignant – given its preeminence with respect to European discovery and settlement; and the subsequent Cushings and Rawlings – and so many others - who loved and continue to love the land, her people and her other species with such great heart. In terms of high biodiversity rarity, artistic traditions, and ruthless devastation, analogies to Hawaii, Alabama, Louisiana, Mississippi, California, Michigan and Wisconsin are not to be ignored. Indeed, the populations of every state in America joined forces to single out all that was wild. In terms of today's bird diversity, California, Texas, New Mexico and Arizona have the four highest levels of biodiversity. All four states are under an abolished umbrella as global warming moves these precious congeries, in desperation, elsewhere, or nowhere.

Protections came, and continue slowly to emerge only after vast swathes of devastation, or at the 11th hour. Our demography follows upon the habitat fragmentation that, in turn, could only take place to begin with by the use of force: the killing of animals.

The Florida panther's plight is more than symbolic: by 1900 Puma [or *felis*] *concolor coryi* was the only big cat left in North America East of the Mississippi River. Centuries of killing the cats up and down the entire eastern half of North America half left this one genetically compromised cat in southwestern Florida at the continuing mercy of major highways, and the second fastest human population explosion in America, after that of California. With "home ranges" that "average about 550 square kilometers for males and 300 square kilometers for females" [*65] the plight of this charismatic creature could not be more severe. It is, along with the Pacific Pocket Mouse and Vancouver Marmot, the most endangered mammal in North America. The panther's remaining territory is no more than 6.5% of all of Florida, and fast disappearing.

In addition to the Pulitzer Prize Winning book by Rawlings, the year 1938 saw two other literary events of the century. The first was that of Jean-Paul Sartre's masterpiece of ecological existentialism, **La Nausée** in which the 30-year old protagonist, Antoine Roquentin, comes face to face with a Chestnut tree, presumably *Castanea sativa*, only to be plunged into a purgatory of metaphysical uncertainties. Sartre's first, and likely finest novel, would serve as the basis of Sartre being awarded – but turning down – the Nobel Prize for Literature in 1964. It is unquestionably the most ecologically-savage portrait of human nature gone awry ever drawn up within urban confines. [*66]

It is "6.00 p.m." and Roquentin describes how he has just been to the park, "sitting, stooping forward, head bowed, alone in front of this black, knotty mass, entirely beastly, which frightened me. Then I had this vision." And he goes on for pages to describe a "veneer," "the diversity of things," "green objects," how "in the midst of a mountain of laurel. All these objects…how can I explain?...The chestnut tree pressed itself against my eyes. Green rust covered it half-way up; the bark, black and swollen, looked like boiled leather. The sound of the water in the Masqueret Fountain sounded in my ears, made a nest there, filled them with sighs; my nostrils overflowed with a green, putrid odour. All things, gently, tenderly, were letting themselves drift into existence…. If you existed, you had to exist all the way… I was In the way for eternity."[*67]

Sartre's early vision in this first of his many fictions, let's us know that conservation, protection, helping others will require a gigantic leap of faith, because all of it – this thing we call ecology; those countless "Others" we propose to assist, assuage, restore, protect – do not necessarily make our own lives any easier. Indeed, our own consciousness can turn against one and all. It requires something other, some deeper laden tool in our genes and convictions and pertinacity that will empower us to succeed at living; to perform the miracles of compassion that existence requires. Sartre rejected "isms" – as in existentialism. But, coupled with ecology, the existentialist purview is perfected as that situation on Earth in which every Being finds itself and must somehow eek out some joy, spread her blessings, manifest a reason to exist.

The third book of 1938 was culminated in the release of an asounding "poem" by Nikos Kazantzakis, **The Odssey: A Modern Sequel**. It was, in some ways, the perfect counterpoint to Sartre's vastly important insights into modern despair, on the very eve of what both Sartre and Kazantzakis knew was about to explode: Germany.

"I said to the almond tree, 'Sister, speak to me of God.' And the almond tree blossomed," Kazantzakis had written in his autobiographical **Report To Greco**, published by Simon & Schuster in 1965 and translated by Pieter A. Bien. This was quintessential Kazantzakis and the event had occurred one cold December day in 1914 upon his descent from a pilgrimage throughout the monasteries at Mount Athos with his friend, the poet Angelos Sikelianos.

The only appropriate summary for Kazantzakis' divine tale – a massively agitated sequel to Homer's original that concludes with Odysseus, hatchet in hand, climbing an Antarctic iceberg - comes from English translator Kimon Friar's extraordinary Introduction to the hefty work two decades later, when Simon & Schuster published the first American edition in 1958.

Outside of this poetic work of genius, Kazantzakis' ecological lyricism is best revealed in a recent anthology of various Kazantzakis selections by Thanasis Maskaleris. [*68] As written in the publisher's overview of Maskaleris' gorgeous little book, "Love supports survival... Nikos Kazantzakis' love of Nature inspired him to write beautiful hymns to Her and to the human life rooted in the soil...having grown up on the fascinating island of Crete — close to trees, animals and wild peasants — he absorbed and retained the terrestrial life in his soul, and made it bloom in brilliant descriptions throughout all of his works. These poetic tributes are not mere 'décor' but a vital source of ever regenerative human life, biological growth, individual spirit and ecological community...Let us hope that an embrace of our natural habitat, such as Kazantzakis summons in each of us, will help galvanize our resolve to respect, revere and collectively protect our one and only home, the Earth." [*69]

Finally, the fourth event of 1938, following upon Rawlings, Sartre and Kazantzakis; an occasion of sheer manipulation goaded on by cowardice and stupidity; a deception of unprecedented consequences was that fateful meeting at the Führerhaus in Munich of Edouard Daladier, Benito Mussolini, Neville Chamberlain and Time Magazine's Man of the Year, Adolph Hitler.

Time Magazine characterized the gathering in Munich on September 29th, 1938, as follows: "When without loss of blood he reduced Czechoslovakia to a German puppet state, forced a drastic revision of Europe's defensive alliances, and won a free hand for himself in Eastern Europe by getting a 'hands-off' promise from powerful Britain (and later France), Adolf Hitler without doubt became 1938's Man of the Year." [*70]

That these four events occurred in one year can have few analogous instances in time where humanity's totality – the best and worst – should show itself.

One runs for cover from this convergence of conspicuous and demonic dualisms. But how does one run from oneself? If, as Sartre intimated, there is a little bit of Hitler in every one of us, how do we transform ourselves meaningfully, in the end? How does human nature accommodate an accelerated accentuation of the Good in us, and a nullifying of the Evil? How do we remake ourselves, become better, if not apotheosized on the altar of our deepest convictions, the St. Francis , as portrayed by Kazantzakis.

Some would argue – and Hugh Lofting did so magnificently, as will be discussed - that inside the mysterious kernel of the eternal child, is a fully-formed adult whose passions are good, realizable, and able to fully defend the rights of all individuals, all species, all modes of consciousness. This is the true conscience, and the child's imagination shows us its location.

Footnotes:

*1 See "Charles Thomson – America's Essential Founder," by John D. MacArthur, Copyright ©2009.

*2 See Heritage Papers, **The Role Of Native Americans In Military Engagements From the 17th Century to the 19th Century**, Edited by Karen L. T. Ackermann, Heritage Books, Inc., Copyright © 2003, Bowie, Maryland, pp.66-67.

*3 Volume One, **Thinking**. Harcourt Brace Jovanovich, Inc., New York, N.Y., 1971, pp.200, 252.

*4 ibid.

*5 **A History of Western Philosophy**, Simon & Schuster, New York, N.Y., 1972.

*6 **Aristotle**, Oxford University Press, New York, N.Y., 1982; See also **The Complete Works Of Aristotle – The Revised Oxford Translation**, Edited by Jonathan Barnes, Princeton University Press, 1984)

*7 "Aristotle biography," http://www-groups.dcs.st-and.ac.uk/~history/Biographies/Aristotle.htm.

*8 See "Last Year In Chalcis, Euboia," in **A Companion to Aristotle**, **Blackwell Companions to Philosophy**, Edited by Georgios Anagnostopoulos, Wiley-Blackwell, Malden, MA, 2009, p.10.

*9 See **The Heretics Feast: A History of Vegetarianism**, by Colin Spencer, University Press of New England, Lebanon, New Hampshire, 1995, cited in the International Vegetarian Union (IVU) website, www.ivu.org/history/greece_rome/aristotle.html.

*10 See "Dining in Ancient Greece –Ancient Greek Food," www.greek-recipe.com/static/ancient/ancient3.html; See also www.focusmm.com/greece/gr_coumn.htm.

*11 See "Analgesia and Anesthesia: Etymology and Literary History of Related Greek Words," by Helen Askitopoulou, Ioana A. Ramoutsaki, and Eleni Konsolaki, Anesthesia & Analgesia – The Gold Standard in Anesthesiology, August 2000, Vol. 91, no. 2, pp.486-491, http://www.anesthesia-analgesia.org/

content/91/2/486.full. See Professor Hele Askitopoulou, Department of Anaesthesiology, University Hospital of Heraklion, Crete.

*12 Anagnostopoulos, ibid., p.13. See Ingemar Düring, **Aristotle in the Ancient Bibliographical Tradition**, Institute of Classical Studies of the University of Göteborg, Sweden, Distributed by Almqvist & Wiksell, Stockholm, 1957, p.462.

*13 ibid., p.11. Citing Werner Jaeger, 1962, pp.320-1 from **Aristotelis Metaphysica**, 1957. See also, http://www.questia.com/search/aristotle-metaphysics for citations pertaining specifically to the various works on Aristotelian metaphysics, as well as the "Aristotle Bibliography," http://facullty.washington.edu/smcohen/433/433bibl.htm.

*14 See **Mankind Evolving –The Evolution of the Human Spec**ies, Yale University Press, New Haven, Connecticut, 1962, pp.132-136.

*15 **Primitive Man As Philosopher**, by Paul Radin, With A Foreword by John Dewey, Second Revised Edition, Dover Publications, New York, N.Y., 1955, p.173.

*16 Harcourt, Brace And Company, New York, N.Y., 1954.

*17 See Lewis Mumford, **In the Name of Sanity**, p.38.

*18 See **Beevor's The Battle for Spain: The Spanish Civil War 1936-1939**, Penguin Group, USA, 2006.

*19 See "25 Years Ago, Heyerdahl Burns 'Tigris' Reed Ship to Protest War," Story by Betty Blair and Bjomar Storfjell, Letter to UN by Thor Heyerdahl, Azerbaijan International, Spring 2003, (11.1) pp.20-21. See www.azer.com/aiweb/categories/magazine/ai111_folder/111_articles/111_heyerdahl_tigris.html.

*20 See **Shamanism: Archaic Techniques of Ecstasy**, Translated from the French by Willard R. Trask, Bollingen Series, Princeton University Press, Bollingen Series LXXVI, Second Printing, Princeton, New Jersey, 1974, pp. 300-301.

*21 See **Poets In A Landscape**, by Gilbert Highet, Alfred Knopf, New York, N.Y., 1957, p.266.

*22 See P.Virgilii Maronis, **Bucolica**, London, Published by Robert Jennings, No. 2 Poultry, **1810**, commencing at page 98.

***23 The Complete Works of John Keats**, Edited by Nathan Haskell Dole, London and Boston, Virtue & Company, Laurel Edition, 1904, Vol. II, p. 113.

*24 **The Rubaiyat of Omar Khayyam, Done Into English by Edward Fitzgerald, With llustratons by Paul McPharlin**, Mount Vernon, The Peter Pauper Press, Stanza 40. No page. No date.

*25 See **The Route Of Parmenides – A Study Of Word, Image, And Argument In The Fragments**, by Alexander P. D. Mourelatos, Yale University Press, New Haven and London, 1970.

*26 See **Parmenides –Fragments and Commentary, Edited and Translated by Arthur Fairbanks in The First Philosophers of Greece**, K. Paul, Trench, Trubner, London, 1898, pp.86-135, Hanover Historical Texts Project, Scanned and proofread by Aaron Gulyas, May 1998, Proofread and pages added by Jonathan Perry, March, 2001, http://history.hanover.edu/texts/presoc/parmends.html.

*27 ibid., pp.88-89.

*28 ibid., p.93.

*29 p.99.

*30 ibid., p.101.

*31 See **Athos –The Holy Mountain**, by Philip Sherrard, The Overlook Press, Woodstock, New York, 1982, p. 159. Sherrard's Notes point to the vast bibliographical underpinnings of what would flourish throughout the Middle Ages and Renaissance in terms of the many writings of the famed **Philokalia on Prayer**, the ascetic works of St Basil, the monks Callistus and Ignatius, **The Theoretikon of the Blessed Theodore**, the **Life of St Euthymios**, the **Life of Peter the Athonite, St Antony**, **Theodore of Studium**, **St Athanasios**, and so forth.

*32 See Richard Ryder, **Animal Revolution: Changing Attitudes Towards Speciesism**, Berg Publishers, Oxford, U.K., 2000, p.17.

*33 op.cit., Mourelatos, p.234.

*34 op.cit., **The First Philosophers of Greece**, by Arthur Fairbanks, p.148.

*35 See **The Mountain Of Names: A History of the Human Family**, by Alex Shoumatoff, Simon & Schuster, New York, N.Y., 1985, p.15.

*36 See "Mysterytopia: Top 20 of the most bizarre creatures," http://mysterytopia.com/2008/09/top-20-of-most-bizarre-creatures-known.html. Blogger Template by B-Themes. See also www.youtube.com/watch?v=blB5neeJHZI.

*37 See **Metamorphosis Insectorum Surinamensium – Insects of Surinam,** by Maria Sibylla Merian, 1705, Taschen Edition by Katharina Schmidt Loske, Köln, Germany, 2009.

*38 See "Scientists Identify Ecuador's Yasuni National Park as one of the most biodiverse places on Earth," Science News, ScienceDaily, www.sciencedaily.com/releases/2010/01/100119133510.htm, Reprinted with editorial adaptations by ScienceDaily from "Global Conservation Significance of Ecuador's Yasuni National Park, PLoS ONE, 2010; 5 (1): e8767 DOI: 10: 1371/journal.pone.008767, by Margot S. Bass, Matt Finer, Clinton N. Jenkins, Holger Kreft, Diego F. Cisneros-Heredia, Shawn F. McCracken, Nigel C. A. Pitman, Peter H. English, Kelly Swing, Gorky Villa, Anthony Di Fiore, Christian C. Voigt, Thomas H. Kunz. University of Texas at Austin (2010, January 19).

*39 See **The Penguin Companion To Classical, Oriental & African Literature**, Edited by D. M. Lang and D. R. Dudley, McGraw-Hill Book Company/Penguin Books Ltd, Harmondsworth, U.K., 1969, p.273.

*40 See Howe's Introduction to **Beyond The Welfare State**, Edited by Irving Howe, Shocken Books, New York, N.Y., 1982, p.3.

*41 See **Tattvartha Sutra – That Which Is, A Classic Jain Manual for Understanding the True Nature of Reality**, by Umāsvāti/Umāsvāmī with the combined commentaries of Umāsvāti/Umāsvāmī, Pūjyapāda and Siddhasenagani, Translated with an Introduction by Nathmal Tatia, With a foreword by L. M. Singhvi and an introduction to the Jaina faith by Padmanabh S. Jaini, The Institute Of Jainology, HarperCollins Publishers, San Francisco, CA, 1994, "5.15, 5.16", p.127.

*42 See **The World of the Japanese Garden From Chinese Origins to Modern Landscape Art**, by Loraine Kuck, with color photographs by Takeji Iwamiya, Weatherhill, New York & Tokyo, 2nd edition, 1980, p.35.

*43 See "Empirical Evidence for Rapid Evolution," by David Reznick, Helen Rodd, and Leonard Nuney, in **Evolutionary Conservation Biology**, Edited by R. Ferrière, U. Dieckmann, and D. Couvet, Cambridge University Press, Cambridge, U.K., 2004, Chapter 6, pp. 101-118.

*44 See "A theory takes wing: Bird feeders not always beneficial," by Jessica Marshall, Discovery News, www.msnbc.msn.com/id/41256806/ns/technology_and_science-science/

*45 See **Prototractatus –An early version of Tractatus Logico-Philosophicus by Ludwig Wittgenstein**, Edited by B. F. McGuinness, T. Nyberg and G. H. von Wright, with a translation by D. F Pears and B. F. McGuinness, an historical introduction by G. H. von Wright and a facsimile of the author's manuscript, Cornell University Press, Ithaca, New York, 1971, "6.1132-6.1251", p.209.

*46 See **Nature And The Victorian Imagination**, Edited by U. C. Knoepflmacher and G. B. Tennyson, Chapter XIV, "A Science against Sciences: Ruskin's Floral Mythology," by Frederick Kirchhoff, University of California Press, Berkeley and Los Angeles, CA, 1977, pp.252, 253.

*47 See **Origins of Nurturance – Developmental, Biological and Cultural Perspectives on Caregiving**, Edited by Alan Fogel and Gail F. Melson, Department of Child Development and Family Studies, Purdue University, Lawrence Erlbaum Associates, Publishers, Hillsdale, New Jersey, 1986.

*48 Multiple private discussions at Ms.Huxley's home. See her non-profit, Our Ultimate Investment (OUI) program, "Teens and Toddlers," www.children-ourinvestment.org/Laura.htm.

*49 See "Green Chimneys: Who We Are," www.greenchimneys.org/

*50 See "Kids & Animals: A Healing Partnership," Written and Directed by Michael Tobias, Animal Planet, www.amazon.com/Kids-Animals-Healing-Partnership-VHS/dp/B00004TWWJ.

*51 Birket Foster's **Pictures of English Landscape**, p.175, Dalziels, XVI.

*52 ibid., XXVII.

*53 ibid., XXX.

*54 ibid., X.

*55 **Landscape Painting – A History**, by Nils Büttner, Translated by Russell Stockman, Abbeville Press Publishers, New York and London, 2000,

*56 ibid., p.175.

*57 ibid., p.261.

*58 ibid., pp.174-175.

*59 ibid., p.372.

*60 http://www.extinctionmemorial.com/specieslists/listFL.htm

*61 **Travels through North and South Carolina, Georgia, East and West Florida, the Cherokee Country, the Extensive Territories of the Muscogulges or Creek Confederacy, and the Country of the Chactaws. Containing an Account of the Soil and Natural Productions of Those Regions; Together with Observations on the Manners of the Indians**, published by James & Johnson, Philadelphia, PA, 1791.

*62 http://www.academia.edu/1432321/The_Tallant_Collection_Early_Spanish_Contact_in_South_Florida

*63 Frank Hamilton Cushing. **Exploration of Ancient Key-Dweller Remains on the Gulf Coast of Florida**, University Press of Florida, 2000, p. X of Introduction. The same sorry syndrome of destruction continues unabated in every sector. The July 19, 2014 front (A) section of the Los Angeles Times, for example, describes two such iterations of this impulse to obfuscate in the name of profit: "Tree tragedy is a painful lesson in Thousand Oaks," by Scott Gold, which summarizes how 170 oak trees were chopped down to make way for an additional "8,000 square feet of space... bringing the total [for an updated shopping center] to roughly 280,000 feet, and new tenants would be welcomed..." (p.A12). Locals, who described a military-like invasion of a flotilla of trucks bearing men with power saws - had always assumed the Oaks – native to California – were protected. It turned out, the laws were ambiguous. While in Bristol Bay Alaska, one of the world's great sites for the spawning of salmon, an open pit mining plan was being debated by locals, the corporation in question and the EPA, pending a month of public hearings. The mine – seeking gold, molybdenum and copper – could allegedly, in a conservative scenario, "fill one of the largest professional football stadiums more than 880 times," with mud and rock, according to the EPA Region 10 Administrator Dennis McLerran. See "EPA curbs threaten Alaska mine," by Maria L. La Ganga, p.A8.

*64 From **Crossing the Creek: The Literary Friendship of Zora Neale Hurston and Marjorie Kinnan Rawlings**, Gainesville: University Press of Florida, 2010, "Marjorie Rawlings" 344, Cited at: http://rawlingssociety.org/rawlings-bio

*65 For a superb summary of the ecology of the Florida panther, and a variety of Strategic Habitat Conservation Area recommendations, see Floridapanther.com at: http://www.floridapanther.com/articles/FLORIDA%20PANTHER.htm

*66 For a more thorough consideration of Sartre in 1938, see **Biotopia**, by Michael Charles Tobias, Zorba Press, Ithaca, New York, 2013.

*67 See **Nausea, A Novel By Jean-Paul Sartre**, Translated From The French By Lloyd Alexander, Hamish Hamilton, London 1962, pp. 170-173 (from the John Lehmann Ltd. edition of 1949 entitled **"The Diary of Antoine Roquentin"**).

*68 **The Terrestrial Gospel of Nikos Kazantzakis**, Translated and Edited by Thanasis Maskaleris, Contributing Essays by Michael Pastore, Michael Charles Tobias, and Jean-Michel Cousteau, Zorba Press, Ithaca, New York, 2011.

*69 http://zorbapress.com/?page_id=168

*70 See: Adolf Hitler: Man of the Year, 1938 - TIME http://content.time.com/time/magazine/article/0,9171,760539,00.html#ixzz2pkN7jy9R

3

THE DOLITTLE CODEX

"The Mysterious Path Beyond"

Wanton destruction is not the normal course of children, amongst any known vertebrates. In humans, children are desperate to create, not kill. Mozart wrote his first opera at the age of 11 ("Apollo et Hyacinthus," K.38, his librettist, an older Benedictine monk from Bavaria, Rufinus Widl, drawing the story and setting from Ovid's **Metamorphoses**). Mozart would continue in a flurry of output, whilst the much older Father Widl would become professor of philosophy a year after his collaboration with the young genius premiered in the Spring of 1767 at the Great Hall of the University of Salzburg.

Percy Shelley was 18 when he wrote his novel **St. Irvyne: or, The Rosicrucian**. Mary Shelley was 19 when she came up with the idea for her eventual novel, **Frankenstein**. Picasso had already displayed his large oil, "The First Communion" in Barcelona by the age of 15. "Bobby" Fischer was 14 when he won the World Chess Championship. Child geniuses learning languages in a day, teens receiving Nobel Peace Prize nominations, children attending universities, becoming surgeons or lawyers, are still the celebrated "exceptions" but, with respect to responding to, and bearing inexpressibly deep connections with nature, it is the rule. An unwavering "telegram" that is answered by the child we all remain, deep down.

The child does not think to engender a children's literature. Rather, the child in all of us does so. These creations are unhampered because they adhere to no other rules than that of nurturance that has been passed down for hundreds of millions of biological years, encompassing unlimited types of individuals across the full biosphere, at every moment in their maturation as persons of profound interest and motivation. Whether one subscribes to theories of fate and destiny, to Furies, nature morte, follies and the make believe, horizons beyond horizons, or doom and gloom with all the futilities wound up tightly in the eternal dark night of the soul, children's literature transcends thousands of years of heady deliberation in an instant, tronies taken from life, but refreshed according to every whim and ideal that is the central pillar of all innocence.

Qualifying all such magical realism, deeply perceived phantasms and wonderfully fantastical renderings – with sure-footed cartography – are the sum total of every fable, allegory, and lesson in life that has the liberty of moving beyond the stilted radars of ratiocination. This is everything, when given to whispering willows or chattering porcelain. Rabbit holes, eyes in the storm, clearing mists, secret gardens, the barely discernable and disheveled professors mumbling revelatory nonsense must take us farther back, or forwards in evolution than the piecemeal, finicky points of the compass we normally carry with us into the dark woods. Indeed, it was the "Fourth Way," said George Ivanovich Gurdjieff that would see us to a true state of consciousness and awareness of self. And, in his 1963 publication of **Meetings With Remarkable Men** Gurdjieff mused on the fact that the only time we truly get lost is when we in fact carry a compass.

Hugh Lofting was once an engineer who relied on compasses, round shank conical drilling bits, valve tools, trenching and other test hammers; all the mechanical trigonometry of mortar, slabtrack, dogspike and hot-rolled steel. The first ancient aqueduct was built in the guise of a 1,036 meter long channel at Samos, designed by the Megarian (Socratic) philosopher/engineer, Eupalinos, long before the Roman military engineers got busy attempting to crisscross half-the-world with their ingenuity. But this was not poetry, and Lofting's heart was not in it.

Later, he found himself a paramedic on the battlefield, at a time, fortunately, when heroine was still generously applied in appeasing the pain of bullet holes in the stomach, and limbs blown hither. He saw the destruction, the needless suffering, of animals, particularly burros - *Equus asinus*, and horses - *Equus ferus caballus*, the sacrificial lambs of cowardly men obsessed with killing one another at all costs.

Then, upon sober reflection, guided, one must conclude, by impulses that would prove to be colossally in tune with a pantheistic reception, he started writing letters to his own children. This epistolary passion nurtured the first of many novels. Novels intended for adults. For the child-like imaginations that adults must never cede to an injurious reality, Lofting endeavored to prove. His proof is in the Puddleby-on-the-Marsh, that epicenter of his mythic English countryside.

"Rural Idylls"

That same rural setting is evident throughout the history of English and Scottish landscape evocations, from the brilliant John Crome of the Norwich School (1768-1821), George Fennell Robson (1788-1833), particularly his breathtaking **Scenery of the Grampian Mountains** (1819), Philip James de Loutherbourg (1740 –1812), most notably his 1822 publication of **The Romantic and Picturesque Scenery of**

England and Wales, or the varied works of Paul Sandby (1731-1809), principally his aquatints of Welsh scenery, and drawings of gorgeous wilderness throughout England and Scotland. By 1764, Thomas Gainsborough had declared that if one wanted "real Views from Nature in this Country" Sandby was the artist to provide them. In his obituary Sandby was called "the father of modern landscape watercolor."

So brief a sketch of English artists verges on the dismissive, if one is seriously intending to even intimate aesthetic justice. How do we dare imagine that bulwark of scenic, zoological, altitude-rich, mist-encinctured, rapturous beastiary, spume-wild, riparian-calmed, Churchyard sedate, heritage-bound, unsullied Kingdom of landscapes and sentience that is the UK; or, as Shakespeare so lavished, a "sceptred isle... This other Eden, demi-paradise, This fortress built by Nature for herself, Against infection and the hand of war... This blessed plot, this earth, this realm, this England (from "King Richard II. Act ii. Sc. 1., of course). Such a world included, at the very least, such artists as William Adam, Joseph Denison Adam, Richard Ansdell, James Aumonier, Thomas Bewick, Henry John Boddington, Richard Parkes Bonington, Samuel Bough, John Brett, Sir Alfred de Breanski, John and William Callow, Thomas Cole, Charles Coleman, Thomas Sydney Cooper, John Constable, John Joseph Cotman, David Cox, James Francis Danby, William Davis, Henry Fuseli, William Havell, Bartram Hiles, Thomas Hill, William Hodges, John William Inchbold, John Knox, George Landseer, Edward Lear, Benjamin William Leader, Henry Stacy Marks, John Martin, John Everett Millais, Edward Moran, Henry Moore, George Morland, Sir William Quiller Orchardson, Samuel Palmer, Helen Beatrix Potter, Thomas Miles Richardson, Jr., David Roberts, Dante Gabriel Rossetti, John Ruskin, Joseph Severn, James Stark, Rev. John Tomson of Duddingston, J. M. W. Turner, and William Witherington.

Imagine so vast a litany of beauty and provocation, lodged in one nation so rich, diverse, and sterling beyond all currency. A setting forth enshrined in its own topography; its own personalities, mood-swings, weathers for every occasion, and five seasons in a day.

This collective had built up the underpinnings of an evocative rural idyll that entranced the young Lofting and, in turn, his good Doctor.

This is no mere exercise, but due praise to that landscape sense and sensibility whose cumulative embodiment we have termed, elsewhere, "the anthropology of conscience," namely, an approach to human cultures historically that have been gently inclined, much about it focusing upon the vegetarian; or have, at the very least, shown a reverence for other species in a manner suggestive of appealing forms of attribution and fair exchange. Justice unstinting; love unconditional.

The landscapes in which this "religion" was received signifies at its most obvious intersection a restraint that cannot but emerge as metaphysical, given the normal human penchant for killing most other creatures and Lofting's strictly contrary example of providing hospice to every mouse, ant, chicken, stray cat and fast-moving brachiator.

The most widely hailed example of such a culture might well be that of ancient Egypt, whose hieroglyphics resound with a veritable chorus of anthropomorphic designations that were all essential to Egyptian self-consciousness, and spirituality (and, it appears, Egyptian vegetarianism).

"Zooglyphic Conscience"

But we see similar zooglyphic evidence from various Southwestern Indian cultures -at Chaco, and among the Chumash and San Nicoleno Indians of California, as well as from the petroglyphs at Winnemucca Nevada, to those previously mentioned (eleven) sites surrounding Burgos, in northern Mexico.

Well prior to all of these images of the interspecies mingling were the varied peoples that coursed the French and Spanish landscapes of the Aurignacian and Magdalenian (Upper Paleolithic) some twenty-five thousand years ago, including Rouffignac and numerous other UNESCO World Heritage sites throughout the Vézère valley coursing the Dordogne.

Not to stop there: note so many similarly reverential artifacts spawned unstintingly across Polynesian, Australasian and African cultures and ideologies; an Indian preoccupation with animal reincarnations, and the role of human/animal relationships among other Near Eastern cultures, like that of the Phrygians with their love of wolves and the cult of the wild child, Cybele, who would, in turn, be inculcated in the Roman goddess of Rhea, linked most provocatively to the Virgin Mary. These connections, without precision but steeped in lyrical association, provide insights into a continuity of adoration that always elicited the child in humanity; a child comfortable in the company of other animals, with an ultimate celebration of the lamb, in Hubert and Jan Van Eck's "Adoration of the Mystic Lamb," the Ghent Alterpiece, completed in 1432 at the Cathedral of Saint Bavo.

Christian, Taoist, Jain and Buddhist mysticism all speaks to interspecies communication. Desert Fathers in Sinai, Taoist immortals along the Nimbo Peninsula, Buddhist painters and poets in Japan, Korea and Tibet, were faithful to the notion of co-habitation with lions, bears, monkeys, birds, and a suite of others. Such wonderful traditions contribute to a long-held advocacy on behalf of the natural world that saw its greatest proliferation during the European Renaissance in such artists as Paulus Potter, Jan Breughel the Elder, Albert Cuyp, and Roelant Savery, painters who depicted an original Garden of Eden.

Shakespeare and Cervantes were certainly part of this cultural journey into the outback. In addition, the rise of illustrated natural history books, from John Latham's late 18th century 7-volume **A General Synopsis of Birds**, to the nearly 40-volume natural history encyclopedia by M. Le Comte Buffon, would excite the graphic revolution in high quality reproductions of nature, that resulted in such acutely sensitive artists as the three greatest painters of birds, Gould, Keulemans and Audubon, wherein science, observation and paradise literature commingled (ignoring for the time being the fact that Audubon thought nothing in his early career of shooting birds for specimen collections, and doubted whether birds had a sense of smell). Of course, their respective painterly powers could be traced back at least 3,500 years to some of the earliest known bird paintings in ancient Egypt; a continuum of deep fascination with birds exhibited by our species in which virtually every year throughout the course of several millennia has given rise to some documented noteworthy artistic and scientific hallmark.

"Moments In The Company of Birds"

To take but a few wonderful examples: 1177, the likely year Abū Ḥamīd bin Abū Bakr Ibrāhīm (c. 1145 – died sometime between 1193-1224), better known by his pen-names Farīd ud-Dīn (الدىن فر ىد) and ʿAṭṭār, composed his exquisite **Manṭiq-uṭ-Ṭayr, The Conference of the Birds**, probably the best known of his more than thirty works, featuring the Hoopoe (*Upupa epops*) considered by the thirty bird species of the world (and the thirty individual birds featured in this epic) to be the wisest of them all, and thus the person to lead them in search of Simorgh, their mythic King. Their odyssey in search of the equivalency of Nirvana lands them on a mountain staring into a lake. There they did not see Simorgh but, rather, their own reflections.

By 1544, William Turner would print a commentary on the many hundreds of birds referenced in the combined works of Aristotle and Pliny. Eleven years later, Conrad Gessner's mesmerizing **Historiae animalium** came out – a gift to the Renaissance, the first true work of zoology ever published, in the form of a 4,5000 page encyclopedia of animal life, four-volumes jettisoned in the robust epicenter of the natural sciences, Zurich, in gorgeous tomes between 1551 and 1558. Gessner focuses on quadrupeds, amphibians, birds, fish, and exquisite snakes. When the German version came out in 1563 under the title **Thierbuch**, Gessner had initiated a royal revolution in perception; turning the Renaissance into a cultural quest for natural history in virtually every household. Every dominion, monarch, donor, patron, political mover/shaker now came to undestand that there was something economically and spiritually at stake in the very Commons and forests that surrounded every village.

With the publication of Pierre Belon's (1517-1564) **Histoire de la nature des Oyseaux** (1555), another milestone was met: the scientific test for the maturation of an internally logical system for classifying avifauna. This science of observation would school the next two hundred years of anecdotal and artistic science, leading to Carl Linnaeus, then Charles Darwin.

In 1757 Carl Linnaeus published a work **On the Migration of Birds** (**Migrationes Avium**), the first work to link climate, agriculture and timing to the movement of birds, as well as seeking empirical and anecdotal information from farmers, scientists and other observers all over Sweden, as well as southern Mediterranean countries (and by inference, Africa). This linking of bird migrations to weather would not be embraced for another century. A year later, 1758, marked the most ambitious taxonomic publication in history: the 10th edition of Linnaeus' world-altering **Systema Naturae**, noting among the 4,400 species of animals – 564 bird species – in addition to 7,700 plant species, each truncated for an elegant, accessible comprehension of Genus and Species.

In 1931, the six-year old Savannah Georgia born Flannery O'Connor (1925-1964) had fallen madly in love with her pet chicken, a buff hued bantam of the Cochin breed. At the height of the Great Depression this little girl and her pet chicken became overnight sensations. Millions of unemployed people worldwide needed that chicken – not to eat, but to be reminded of something… Thirty years later in her essay "Living With A Peacock" (Holiday Magazine, September 1961) O'Connor described how Pathé News had heard about this chicken of hers who could walk backwards. Pathé sent a cinematographer to record the chicken genius (available as a two-and-a-half-minute YouTube). O'Connor described the moment as a "highlight" in her life, everything thereafter, an "anti-climax." In the opening paragraph of her essay, she also wrote (in a manner bearing the same astonishing irony and insights she brought to bear on her two extraordinary novels, **Wise Blood** (1952) and **The Violent Bear It Away** (1960), "I suppose there was nowhere left for her to go [referring to that chicken] —forward or backward. Shortly after that she died, as now seems fitting." O'Connor also elaborated on how she would sew for her chickens, including "a white piqué coat with a lace collar and two buttons in the back" for a bantam she called Colonel Eggbert.

Late that same year that O'Connor's chicken became an international star walking backwards, just a few days before Christmas, author E. B. White's dog Daisey died, struck by a taxi cab. Several months later in the New Yorker Magazine, March 12, 1932, White immortalized Daisey in a lyrical, insightful, utterly human-like obituary, concluding, "She never grew up, and she never took pains to discover, conclusively, the things that might have diminished her curiosity and spoiled her taste. She died sniffing life, and enjoying it." [*1]

Around the same time, Paddy, a wolf dog who hung out with the longshoremen in Wellington, New Zealand's harbor, also passed away, received an obituary in the national press, and at Paddy's funeral, hundreds of mourners arrived to say their deeply affecting fair-wells. [*2]

For O'Connor, who also communed with scores of other birds and countless other species, her Puddleby-on-the-Marsh, as Hugh Lofting had called his own version of paradise, happened to be her Uncle's dairy farm, named Andalucia, on the edge of

Milledgeville, Georgia, a sizeable pocket of heaven that today comprises a 544-acre estate under the loving aegis of the Flannery O'Connor - Andalusia Foundation, Inc. With Tobler Creek running through this wild central Georgian mini-wilderness, there exists a fine emblematic suite of biota and species characteristic of Southeastern America, with its raptors, beavers, hardwood (mostly oak) ecosystems, bogs, pasture, marshland and exquisite members of the Phasianinae sub-family of Common Peafowl milling about, *Pavo cristatus*, peahens and peacocks. During her all too brief lifetime, O'Connor had lived with dozens of them (an alleged number of 50 at one time). Benjamin Ashworth's **A Field Guide To The Common Birds of Andalusia – 40 Birds: Their Lives And Influence On The Works of Flannery O'Connor** [*3] includes such treasures as the Barred owl and Belted kingfisher, Brown-headed nuthatch, Georgia's State bird, the Brown thrasher, the "Carolina chickadee, Carolina wren, Chipping sparrow, Downy woodpecker, Eastern bluebird, Eastern kingbird, Eastern meadowlark"… "Eastern towhee," "Great crested flycatcher," in addition to "Pine siskins," "Ruby-crowned kinglets," "Tufted titmouse," "Northern parula," "Whip-poor-wills," "Yellow-rumped warblers," "White-eyed vireos," and "Red-bellied woodpeckers," among deciduous scrub, pine and mixed forests and bottom-land swamp, marsh, open canopy forest, fields and forest edges.

These biomes have been much transformed with human encroachment since the time of William Bartram's **Travels** (1791) throughout Georgia and elsewhere in Southeastern America. Species names and Genera have been re-assigned. Some species have vanished, or greatly diminished in number, as everywhere else in North America. But as Edward J. Cashin of Augusta State University poignantly wrote in 2004 in the **New Georgia Encyclopedia** [*4] in an entry entitled "William Bartram in Georgia," "Bartram wrote his book while the founding fathers drafted a constitution for the republic. He saw divine guidance at work in the shaping of the new country. 'I foresee a magnificent structure and would be instrumental in its advancement,' he wrote. His book exalted the human spirit and celebrated the richness of America's natural world. He called upon Americans to respect the rights of Indians, to eradicate slavery, and to live up to the best in themselves.'"

"The Search for John Hunter, Esquire"

Explorers and literary savants of the 17th and 18th centuries advanced the case for the dignity, communication skills, and psychological depth of other species by their (sometimes fictional, at other times acutely detailed) studies. William Gilpin's (1724-1804) notes on the gregarious and communicative disposition of British wild bores in his 2-volume **Remarks On Forest Scenery** (London, 1791) could easily have been written by Hugh Lofting. Similarly, the images of the great nature painters mentioned above, particularly those of the aforementioned John Crome of Norwich provide all the 19th century allusions and symbolic pastoralist images that would form the ethos for Dolittle's eternal rural England.

The mysterious Dr. John Hunter, and the irrepressible botanist, Erasmus Darwin furthered that rural renaissance of discovery.

Joseph Adams' **Memoirs Of The Life And Doctrines Of The Late John Hunter, Esq.** (London 1817) suggests rather remarkable similarities to Hugh Lofting's early years in Maidenhead, Berkshire. Hunter was a veterinarian, and perhaps the first horse whisperer in Western tradition to have been described. It is possible that Lofting based his Doctor, at least in part, upon this fascinating individual who lived not thirty miles from where Lofting would be born.

Adams' recounting of Hunter's life centers upon the veterinarian's prodigious publications about, and observations upon animals, in addition to his human medical research. During the height of his career, Dr. Hunter settled near Leicester Square, on Castle Street, his nearly 3,500 specimens becoming part of the Hunterian Collection of London's Royal College of Surgeons at 35 Lincoln's Inn Fields.

Adams described Hunter's passion for natural history, his time at the Bencoolen settlement in Sumatra as assistant-surgeon for the East India Company where he obtained a double-horned Rhinoceros and an "uncommonly formed fish". He studied snakes, and authored a paper on "The Heat of Animals and Vegetables." Upon returning to London he purchased a lease hold for twenty-four years on the east side of Leicester Square, a large house in which he assembled his already considerable collection.

Not content to become a curator of his own dead things, he applied his considerable talents towards a rethinking of Aristotle and the pursuit of true veterinary experiments as part of the newly formed Veterinary College of the Agricultural Society at Odiham, in Hampshire. To this end, and by which his posterity would become assured, he befriended "less known kinds of bees" as well as silk-worms and their eggs. And it became, as Adams wrote, "a favourite amusement in his walks, to attend to their actions and their habits, and to make them familiar...The fiercer animals were those to which he was most partial, and he had several of the bull kind from different parts of the world. Among these was a beautiful small bull he had received from the Queen, with which he used to wrestle in play..." His house became a true retreat in which all creatures – every member of the "bee tribe" particularly – were welcome. Bulls, birds, wasps, a poultry-yard in back teeming with his good friends; a "pleasure ground" for both scholarship and the first ever life-long experiment in interspecies recognition and joy.

Hunter would write his famed **Animal OEconomy** and is said to have viewed every operation of Nature "as standing miracles, evincing a first incomprehensible and all-powerful cause..." He was in love with birds and would demonstrate how the "want of stones" would cause them no end of misery. Even a button would suffice, in a stone throw's absence.

His myriad writings included an essay "Of certain Receptacles of Air in Birds,"(1774); "On the Gillaroo Trout" and "the Gymnotus" that same year; "A Natural History of Teeth"(1778), of "Free Martin"(1779), a year later, an "Account of an extraordinary Pheasant," followed two years later by an "Account of the Organ of Hearing in Fishes" (particularly the lion and sea-cow, in which hearing was believed to be far more advanced than in man). In 1785, "Anatomical Remarks on a new Marine Animal"; in 1787 "Observations tending to show that the Wolf, Jackall, and Dog, are of the same species". Later that same year, "Observations on the Structure and OEconomy of Whales" and subsequent research papers on bees, jackals, wolves, more dogs, and a remarkable piece "On a Secretion in the Crop of breeding Pigeons for the Nourishment of their Young, " research carried out – not at Trafalgar Square, which did not exist in his day, occupied instead by Royal Households - but across the much-relished territory of Leicester Square in his own neighborhood.

"The Renaissance of Leicester Square"

And what a neighborhood! Doors away was the famed Holophusikon Museum, or Leverian, as it was also named, after its owner, Ashton Lever, the one time High Sheriff of Lancashire, who would develop a passion for animals, collecting 28,000 for his own enormous museum – the largest of its kind at that time in Great Britain.

Lever's multi-storied cabinet of natural curiosities was expanded when the admiring Captain Cook donated objects from his diverse voyages, and this entire edifice of natural history would be exhibited at Leicester House, near to the residence of Frederick, Prince of Wales. The liveliness of the atmosphere; the very celebrity in the heart of London, further galvanized the good Dr. Hunter's passion for such creatures as his friends, the everyday pigeons, but also the far more exotic hippopotamus, tapir, argus-pheasant, guanica and camelopardos, as he called them.

A catalogue of these diverse newcomers to Great Britain was produced by the celebrated George Shaw, best remembered for his massive 16-volume **General Zoology, or Systematic Natural History** (1802).

In 1786, the lack of revenue induced Lever to sell his collection to a James Parkinson, the discoverer of Parkinson's Disease, who would go on to write a famed treatise in 1804, **Organic Remains of a Former World**, and describe the 7 days of Creation (the Heptameron, plus Day 7) as governed by a quite novel chronology, such that each day increased in hours, the Day of Rest accounting for 10,000 years. He also invoked Noah's Ark (doubtless influenced by the goings-on at Leicester Square) and propounded the belief that extinction was indeed a fact made manifest by acts of God.

By 1788 Parkinson moved the entire collection to the Rotunda at No. 3 Blackfriars Road, and two years later, by an Act of Parliament, sold off every creature, fossil and other member of the museum at public auction. Among the buyers was a 20-year old Irish zoologist, Edward Donovan who would go on to write **The natural history of British insects explaining them in their several states, with the periods of their transformations, their food, oeconomy, &c. together with the history of such minute insects as require investigation by the microscope, the whole illustrated by coloured figures, designed and executed from living specimens** (London Printed for the author, and for F. and C. Rivington, 1792-1813, 16 volumes). In addition, such was the busy fanfare emanating from this bustling ecological epicenter that was Leicester Square, Donovan went on to explore and write up phenomenal accounts of creatures and their biomes ranging from New Guinea to New Zealand, even **An Epitome of the Natural History of the Insects of China**, and similar works about India and New Holland (a nation of many nations.)

Portions of Parkinson's collection were also acquired by the Imperial Museum in Vienna, and by Edward Smith-Stanley, 13th Earl of Derby who installed 1,272 of the still living community of animals at his Knowsley Hall near Liverpool. He supported the great ornithologist and natural history painter Edward Lear and would be honored by his name being applied to *Psittacula derbiana*, the Derbyan (or Lord Derby's) parakeet. [*5]

"The Derbyan"

That particular parakeet has its own story and geography of no small importance, or spectacular beauty. It would be formally introduced to science in 1849 by Louis Fraser, author/publisher of **Zoologica typica: or Figures of new and rare mammals and birds described in the Proceedings, or exhibited in the collections of the Zoological Society of London.** [*6] and Curator of the Museum of the Zoological Society of London, from which ultimately would emanate one of the world's first activist charities focused on conservation biology and the protection of species around the world.

The precise status of this Psittacine (true parrot) from the border area between Arunachal Pradesh (India) and China, is currently uncertain, though the IUCN (International Union for the Conservation of Nature (formerly, ...of Nature and Natural Resources, or World Conservation Union) considers it of "Least Concern". But as recently as 1989, it was being "heavily trapped in Sichuan and East Tibet and (was) probably rapidly heading for extinction." [*7]

Because it is a monotypic species (its gene flow undifferentiated amongst numerous populations that all come within the same species designation), the Derbyan parakeet's range may well encompass most of the Eastern Himalaya, where it loves socializing in alpine rhododendron forests, making it one of the few mountain parrots in the world, along with New Zealand's Kea. The Derbyan parakeet's apparent friendliness in captivity has given rise to much abuse of this extraordinary bird. The IUCN recognizes 19 known modern (post-1600) Psittacine extinctions, with at least 130 of the more than 350 known parrots listed as Threatened, Endangered or Critically Endangered. In at least one case (a lorikeet endemic to New Caledonia) no individual has been found in over a century, but it has not yet been deemed extinct. [*8]

The aforementioned Fraser had taken part on an expedition down the Niger River, became a political Consul at Quidah, Dahomey (later to become Benin, in West Africa), continued on his journeys in 1859 to Ecuador on behalf of the English lawyer, zoologist Philip Lutley Sclater (1829 -1913) who would become the President for the British Association for the Advancement of Science, discovering new taxa on continent after continent, and amassing some 9,000 bird species that he would donate to the British Museum.

Fraser himself would be honoured with at least three creatures named after him, an Eagle-Owl, a Warbler and a Musk Shrew. By the time Fraser was in Ecuador collecting birds and mammals for his benefactor, that same Sclater had recently published his belief system focusing upon so-called Lemuria, or Lost Lands (a "lost world"). Sclater sponsored not just Fraser, but multiple scientific adventures from Mexico and Madagascar to India and Ethiopia and his publications helped fuel the greatest naturalist zeal ever witnessed in a single generation – again, all streaming from this now mystical Leicester Square. His own work was vast, including research on antelopes, and ornithological collaborations with Osbert Salvin. He also joined forces with the Argentine born scientist and author, W. H. Hudson, a founding member of the Royal Society for the Protection of Birds. Hudson is perhaps best remembered for his fiction and non-fiction, including **Green Mansions** (1904), **The Purple Land** (1885) - referenced in Hemmingway's **The Sun Also Rises; Far Away and Long Ago** (1918), and his most ambitious work, the magnificent two-volume Argentine ornithological **The Birds of La Plata** (1920). [*9]

Seven species would be named after Sclater, including a blue-eyed Lemur (which the Authors herewith were among the first to film at a remote German research camp {somewhere} in northern Madagascar), a Parrotlet, Monal, Penguin, Cacique, Chickadee and Cotinga. [*10]

As for Salvin, he would travel throughout Central America, became Editor of The Ibis, the most famous journal of ornithology, and accepted the post of Curator of the Strickland Collection of ornithological specimens at the University of Cambridge. [*11]

We don't know if Dr. Hunter himself ever acquired any of the original creatures that occupied the peculiar cornerstone of global scientific studies that was Ashton Lever's original museum, but we can state with some confidence that Hunter's time and place, his personal passions and unrelenting friendship to animals coincided with a most remarkable moment in the history of science and inter-species cohabitation.

His late 18th century passion does much to illuminate his successor, Hugh Lofting, and the character of Dr. Dolittle. This protagonist of the most beloved "children's" classics of all time had the full weight of the previously intimated English landscape – beginning with Chaucer, then Shakespeare – to bolster his case for an amusing fellow of some obscure charm, brilliant medical instincts but utter absent-mindedness when it came to being a suitable householder, or practical explorer – in the "adult" sense (since he had no qualms about planning space travel without a rocket ship, or journeys to Africa on the backs of larger sea roaming creatures). A diplomat of every biological kingdom, plus some, whose portfolio brimmed with the prowess of a young boy discovering the world for the first time (absent the grim philosophical musings of a Blaise Pascal's **Penseés**, or T. S. Eliot's **The Wasteland**); of a Gentleman in an apron, a scientist in command of play money; a genius without periphery. In sum, a good soul who had quite irreversibly fallen in love with all animals, and made this a condition of the social contract in any neighborhood, on any given day. A foremost attribute, in other words, of any decent, Universal Law-abiding Englishman and of all civilizations.

"The Genesis of Dr. Dolittle"

As indicated above, Lofting witnessed the pain-wrenching death of horses in the "War Horse"-like trenches of World War I. Seeing horses put down, while soldiers were surgically treated, then healed, left a mark on his sense of fairness in the world that gives us much to think about when contemplating the bedside manner of Doctor Dolittle and the obvious link to Dr. John Hunter's own apparent dilemma: medical expertise and a near universal indifference by his species to other creatures, "specimens," at their best.

Those exotic animals shipped in to the aforementioned collections and museums, dead or alive, are an historic metaphor for most natural history museums and their current dilemma, a crisis of message and emotion most pronounced and expressive in the large body of Dr. Dolittle books, namely, how to convert our fascination with the natural world into a proactive love affair with life, not with death. To restore and re-wild wilderness, not fabricate zoos for the edification of a species given to callous and imbecilic voyeurism.

By what means, however obvious yet least practiced, to engage in conversation, not sarcasm; to be enlisted and empowered by beauty, not jealous of it; to feel not at odds with that we cannot understand, but rather, humbled and delighted with the mystery of the Creation. To embrace the unknown in and of itself; neither adding nor subtracting; not for calculation or for gain, but in sheer joy before the clouds of unknowing.

Dr. Dolittle does all that – giving himself over to his adventures, to all the animals who inhabit his universe. They seem to have originated not just from one author's imagination, but from a history of associations at whose matrix lies a late 18th-century ensemble of extraordinary goings-on between the pursuit of scientific knowledge, a frenzy to know the world in all of its far-flung geographical quadrants, but also to transform the paradigm of natural history such that an ethological physiolatry might be apotheosized into the living hub, a veritable Gare du Nord of our relations with other species.

In essence, Lofting took a history of rural England, the British rage for natural science, with all of its Darwinian arguments and controversies, and amalgamated them into a singular orientation that would become the hallmark of animal rights.

"A Profound Discussion Amongst Flowers"

There were other forebears of this revolution, to be sure, Erasmus Darwin among them, Charles Darwin's celebrated grandfather whose remarkable Poem in Two Parts, **The Botanic Garden –The Economy Of Vegetation, and The Loves Of The Plants, With Philosophical Notes** (London, 1791) gave "new echoes to the throne of Love" and argued in its First Canto (the entire work illustrated and written as poetry) to "The Genius of the place (that) invites the Goddess of Botany." Should he be credited with the theory of Evolution? Certainly Erasmus' devotion to Love as the crux of biology gives him a uniquely inspired vantage point in the history of humanity's tug-a-war with nature. He did not forsake a single blade of grass in his philosophy, attempting instead to revive the most ancient and animistic reverences for those things largely ignored or obfuscated by the more manifest rage to collect oddities and "deceased" representatives of every living being, rather than befriending them.

Keeping in mind today's frantic efforts by cyber-taxonomists to catalogue the entire colossus of genes and genomes; to store the Earth's biomass in submerged Norwegian arks or government and university cryogenic labs lest we find ourselves marooned on this planet sometime soon, bereft of all those biochemical allies that keep us alive, not least of which the bacteria in our bodies, men like Erasmus Darwin and John Hunter give loft to the urgency of a Hugh Lofting, whose books, the so-called children's books, are nothing less than the blueprint for all children of the future, namely, humanity itself if it is to survive. This should be more self-evident than any other Declaration.

Aristotle (384-322 BCE), who knew eight groups, and named about 170 bird types, anticipating Saint Francis' (1181-1226) own conversations with birds, irrespective of their numbers or types - Francis did not, from what we can discern, discriminate amongst his Fellow Beings – gave evidence of the human love of nature that Lofting's Doctor enshrines. That the Doctor is a scientist, of sorts, as well as a pantheist, places him squarely in that span of contradictions which encompass a Mahavira, an Aristotle and a St. Francis, all in one personage.

Alfred Wallace presupposed that the sheer beauty of feathers (note the great "Wing of a European Roller," by Albrecht Dürer, 1512,) and of the birds they adorn, speaking specifically in this case of the Aru Islands of New Guinea and the Great Bird of Paradise, were surrounded by "a race of the most savage and ruthless character in the very lowest stage of civilization. In such a country and among such a people," he went on (unconcerned with any posterity ingratiating those who would surely read him) are nonetheless "found these wonderful productions of nature. In those trackless wilds do they display that exquisite beauty and that marvelous development of plumage, calculated to excite admiration and astonishment among the most civilized and most intellectual races of man. A feather is itself a wonderful and beautiful thing…" [*12] Note his emphasis upon the word "calculated". What are we to glean from this insight, if not an extraordinary intelligence, unequivocal sophistication, despite his serious malaria inhibiting what might have been an even more exalted and intellectually unprecedented poetic.

This was in accord with those non-human "higher" states of being, all previously subscribed to by poets like Erasmus Darwin, and William Gilpin's groundbreaking works in which he outright declared (as would the Belgian poet, philosopher and playwright, the 1911 Nobel Prize winning Maurice Polydore Marie Bernard Comte Maeterlink [1862-1949] in his astounding book of philosophical science and emotion, **Hours Of Gladness** [*13], that plants exhibit qualities of love. Moreover, many of the late Enlightenment painters and explorers and naturalists made an art form of immersing themselves in the realms of other species. Alexander von Humboldt not infrequently fell into a kind of interspecies trance before the great monuments of nature that he added matter-of-factly to the map of South America during his many years of exploration. Many philosophers were actually taking the time to learn the language of different species, to the extent they thought it was possible. At the same time (late 18th and 19th centuries) there were the first breakthroughs in legal debate concerning anti-cruelty laws, from Norfolk Island in the Pacific to England.

Such efforts would lead to the examples of John Muir and D.H.(sic) Thoreau, and the invitation to believe in the immanence of interspecies relationships as figured in some of the greatest of the Romantic, Hudson River School, Impressionist and Luminist painters in North America and Europe. The most obvious names, of course, are Albert Bierstadt (deer and bear), and Thomas Moran, who liberated in human consciousness, with his great art forms, the life force in mountains and waterfalls and the very storm clouds over the Grand Canyon. As with the most majestic of the myriad man who was Bierstadt, a veritable Taoism can be instantly detected in that spectacular series of watercolors and chromolithographs Moran (1837-1926) produced upon visiting the many thermal vents and calcified geologic formations of Yellowstone: his book (a copy of which, today, can be had for roughly half-million dollars) **The Yellowstone National Park and the Mountain Regions of Portions of Idaho, Nevada, Colorado and Utah. Described by Prof. F. V. Hayden. Illustrated by Chromolithographic Reproductions of Water-coulour Sketches, by Thomas Moran**, Boston: L. Prang and Co., 1876. Folio. (22 x 19 inches).

The human/animal/landscape communications are no less prolific in George Catlin (bison and Native Americans); Martin Heade (orchids and hummingbirds); and Edward Church (the life of the tropics).

These inventions had long been anticipated, by the Mother Goose tales, and the unfolding revolution throughout the realm of all subsequent Aesop-indebted children's books involving animals and their interaction with people: E. Thompson Seton, Rusticus, Lewis Carroll's Alice Lidell, Beatrix Potter's **The Tale of Peter Rabbit**, and A.A. Milne's **Winnie-the-Pooh**. No examination of this material would be complete, of course, without L. Frank Baum's **Wonderful Wizard of Oz** (if one can get beyond one of Baum's incendiary epistolary remarks regarding American Indians); and most assuredly, Kenneth Grahame's **The Wind and the Willows**. These may be thought of as the iconic emissaries of something far more richly veined in the substratum of the human psyche than mere "genre". These were Sistine Chapels; Restoration Literature; Reformation theologies; Post-Modernist sensibility; Marxist manifestos, if you will, giving all of Nature and her progeny the Rights conferred upon men and women. Beasts were no longer to be burdened, but welcomed as part of a great family.

"The Author's Peregrinations"

Lofting had first developed his Dolittle character whilst sitting in his tent at night on the frontlines of World War I in the trenches of Flanders. Prior to the war, he had moved to New York and was working for the British Ministry of Information. He'd married Flora Small (a New Yorker), and his first child, Elizabeth, had been born in 1913. His son, Colin was born in 1916. Driven by a restless nature that had first expressed itself in his earliest travels to West Africa and Cuba, as far back as 1904, Lofting enlisted, and became a captain of the Irish Guards. Confronted by all manner of horror in the "War to End All Wars" (a slogan that would later haunt his works as he developed his anti-war sentiments within the body of the Dolittle sagas themselves) he took on the attitude of the flawed but sterling Emperor and philosopher, Marcus Aurelius, who of course wrote his own ***Meditations*** from the battlefields. **The Story of Doctor Dolittle** began as letters to Lofting's two children that took as their fuel all that was opposite the war in human nature. He sought his own paradise that logically flowed from the earliest memories of England, and - we may surmise - his wide reading of the arts and sciences, literature and history. When Lofting was injured by shrapnel to his thigh, he repaired to England where Flora and children joined him, and he took up his writing with a greater zeal and purpose in mind, it may be divined.

Returning to the U.S. in 1919 - as described masterfully by biographical critic Gary Schmidt - Lofting met the British writer Cecil Roberts on board a ship, and was encouraged to seek a publisher for his tales. Frederick A. Stokes was the man recommended, and his company's editor, Helen Dean Fish. A correspondence developed, as Lofting moved his family to a 90-year old farmhouse on Chestnut Hill Road in Wilmot Connecticut, a town steeped in historic tales of violent encounters between residents and the local wildlife, particularly panthers and black bears. He would work in the two-story barn outback, looking towards the woods.

Later, the Loftings moved to Madison, Connecticut. **The Story of Doctor Dolittle** was published in 1920 and its inhabitants, their integral connections, scenes of unfolding behavioral oddity, sublime attitudes and underlying amusement, became an instant source of celebrity throughout England and the U.S. By the 1960s, there would be more than sixty editions that had been released just in the United States. Such was the fuel of this fanfare giving Hugh Lofting his full-time provenance and no better obsession. While the dozen Dolittle books which would preoccupy his next two decades were the core of his literary output, other works would emerge, as well as poetry, and some non-fiction article writing. As early as 1924 he published in the Nation an essay appealing to a non-violent conscience.

"The Discovery of Puddleby-on-the-Marsh"

What was it that so transformed Hugh Lofting into an animal rights pioneer and illustrator? Like such legends as "Brigadoon," "El Dorado," and "Erewhon," this Puddleby-on-the-Marsh, about 25 miles from London, became, for Lofting, the center of Dolittle's universe. It embodies what Ralph Waldo Emerson describes as the perfect balance between the wilderness and civilization; libraries and pearly streams. And, as we know, a world where animals discuss all the problems of the neighborhood and of animal nature; they gossip, and complain to the good Doctor. It is a logical, reasonable, all-intuitive universe where discourse is witty, appealing, compassionate and infectious; like any fine Irish pub where all the problems of the world are solved; and where great adventures happen at the spur of the moment, and concerns about money, a roof over one's head, play second fiddle to the greater aspirations - the saving of a species of monkeys in Africa, or abstruse problem solving among mice, cats, hogs, and horses. Lofting's world somehow provides a bulwark against Wordsworth's "world too much with us" - with our tens-of-millions of pets; and a century of environmental activism, and animal rights, and ecological disasters. Little wonder that the paradigm that motivates Dr. Dolittle emerges at a time when antivivisection in the U.K. had achieved trend-setting results in terms of public policy.

Equally, if not more important for the purposes of his Puddleby Capital, Lofting's own personal world had changed remarkably, so that he could transport his readers into a familiar language and landscape, away from the horrors of World War I and the failed Treaty of Versailles. He wrote in a manner that lent nostalgia to this peopled planet, and gave us hope; not blissed-out hippy prospects in modular tepees, but a new currency for feeling good about our ability to make a difference here, at once, right in our front yards. His initial outpouring of books (about one per year) provided readers of all ages and persuasion something to smile about. In literary terms, his conceits proved to be unique, and Lofting - the writer, the man - was duly rewarded.

Puddleby-on-the-Marsh remains enshrined as an ecological beacon of moral and ecological sanity for the 21st century, largely on account of the accumulated weight of British rural zoology, farming traditions, and the intersection of all those garden and architectural design elements that were modeled after a vision of nature. Much of that arabesque of natural elements derived from those designs by William Kent (1685-1748) and Robert Adam – the penultimate Georgian visionary – as well as the luscious estates made for lovers and love lost in the world of Jane Austen; of Capability Brown and the Regency aesthetics of landscape architect Humphry Repton (1752-1818).

In Repton one sees the ultimate precursor for admixing bowling greens and the science of hortoculture, with imaginative experiments, even sensual hiding places. We see these components scattered throughout his considerable body of work. From his **Sketches and Hints on Landscape Gardening** (1794), **An Inquiry into the Changes of Taste in Landscape Gardening, with Some Observations on its Theory and Practice** (1806), his **Designs for the Pavilion at Brighton** (1808), to his many famed "Red Books" (beautifully bound red morocco tomes with gold stamping) containing his hand-sketched and colored images and methodical advice he would produce for potential clients who desired landscaping at their country homes – from Norfolk, Somerset, Ipswich and the great romantic, foliage-wild kingdoms of Blaise Castle, to Stoneleigh and Woburn Abbeys, the former with its Chinese garden, to the principally dignified and user-friendly social gardens as those embodied in Russell Square in London, commissioned by the Duke of Bedford. Repton also composed a very practical but also luscious essay for the Transactions of the Linnean Society, xi. 27, "On the supposed Effect of Ivy upon Trees."

So there was this immediate tradition of follies and miniature wilderness worlds-within-worlds teeming across the imagination of Puddleby; fantasy gardens like those envisioned upon a stage by Scottish playwright and novelist, J. M. Barrie. Sacrosanct habitation for Beatrix Potter's "soporific flopsy bunnies," of the swans at Kew Gardens and the French painterly phantasms of Jean-Honoré Fragonard, Hubert Robert, Nicolas Lancret and Jean Antoine Watteau. Even the great Dutch 17th century moments of village quiescence can be likened to the Doctor's household for it commended an eternal pandemonium paradise of communing individuals more or less at peace beneath the grand umbrella of biophilia, as also witnessed in the many delirious ice-skating rinks of the Dutch 17th and 18th century pictorial winters. The prolific and ironically mute Hendrick Avercamp's (1585 -1634) "A Scene on the Ice Near Town" or "Winter Scene on a Canal" and "Fun On the Ice" come immediately to mind.

In Dolittle's world this precursor of an historic garden ethos is all about its inhabitants. It is a fact that Dolittle inimitably drives home and is important to recognize in an era (today) given to the ideal of wilderness and habitat protection but without any particular reference to individuals therein. In Dolittle we share intimately with the Avercamp-like density of memorable personae who indeed are clear beneficiaries of the sanctuary, as defined by the earliest church and cathedral sanctuary policies, normally afforded political or other refugees in England. Both King Aethelberht (ca.AD 600) and the pre-Saxon King Dunvallo Molutius (4th/5th centuries BCE) are said to have instituted political asylum within churches, a system divided between lower, unsanctioned asylum, and those sanctuaries actually licensed by a Norman authority. James I in 1623 abolished all of these routes of safety but even a King could not disavow that which spiritually is incarnated across the whole landscape of religions to this day. Every Church, for example, has its sense of God being present in the Eucharist; of the Protestant chancel; of altar rails, antimensions, holy tabernacles, Ark of Covenants, altar-tables, naves, Holy of Holies, iconostases (walls of icons) and the like. These physical forms within a temple all comport with an ecological sensibility that is the garden as employed by Lofting's universal genius in describing the household of Dr. Dolittle.

And the household companions? Polynesia, the parrot, of course, who first introduced the Good Doctor to animal communication; Dab-Dab, Dolittle's housekeeper (a duck); Jip, the dog who assembled the Home for Crossbred Dogs; Chee-Chee, the Doctor's African Tracker; pushmi-pullyu, the two-headed equine of absolute efficiency and decorum; Cheapside, the London sparrow; and, last but surely not least, little Tommy Stubbins (a human lad), Dolittle's chief assistant.

In one particularly unusual sequence, figured in his 1927 novel, **Doctor Dolittle's Garden**, the Tenth Chapter focuses upon "The Hermit Dog," who finds a life in the wild attractive, at least for the first few seasons. "So I was now going to live independently of him," says the dog, speaking of humans. "I think part of the idea was proving to myself, as well as to 'Mankind', that a dog could be really self-supporting." [*14]. This same dog had been a "tramp dog, a performing dog, a gipsy dog, a monk dog, a professional sheep herder" and now would become "a hermit wild dog." But in the end, he and an eventual fellow companion dog find the company of humans too much to do without. They remember respectively all the good times they'd had with children on farms, and have heard of this magical place, where dogs can live totally free to be themselves, but in the company of humans, at a village called Puddleby-on-the-Marsh. [*15]

What a twist on the animal rights ethos. Moreover, science has begun to catch up with the creatures of Lofting's imagination. Studies of sheep, and passenger pigeons, of parrots and dogs now unambiguously reveal a level of neurological, emotional, and cognitive know-how that was never before understood, at least until Lofting. What is so compelling, of course, is the ease with which Lofting submerses us in the world of interspecies communication. Both children and adults accept that dual-nature without question. It feels right. It is the ambience we all desire, or many of us do.

"Later Years"

Lofting's controversial anti-war sentiments and articles, particularly his extended poem -"Victory for the Slain" (London: Jonathan Cape, 1942) formed the backdrop for his final years in Topanga, California, a wild vegetated moist coastal gorge between Santa Monica and Malibu, with a Mediterranean floristic province composition of chaparral, woody and evergreen shrubs, Fremont's cottonwood, live oaks and California sycamore, even relic species from the last Ice Age which the Santa Monica Conservancy (LAMountains.com) references as "white alder, big-leaf maple, and black cottonwood." The Conservancy also points out that "over ninety-five percent of the native grasslands have been displaced by foreign invasive plants." This would have been the case, with little to differentiate the transitions, during Lofting's time in Topanga. As for animal friends, Lofting and his family co-habited Topanga with what the Conservancy lists as "coyotes, mule deer, bobcats, raccoons, and skunks... Just away from the urban edge, other predators, such as grey fox, mountain lion, American badger, long-tailed weasel, and ring-tailed cat, [who] occupy various niches." In addition to "numerous prey species—such as rabbits, squirrels, rats, mice, and other rodent species. Seven species of hawks, eight species of owls, peregrine falcons, golden eagles, northern harriers, American kestrels, and white-tailed kite share in this bounty of prey... [and] "over eighteen species of snakes and eight species of lizards in the zone." [*16]

It was during this late period in his life that the now celebrated Lofting veritably disappeared off the biographical radar screen, despite his great fame. He was not part of the Hollywood phenomenon (no efforts to write screenplays or take up famous exile, like a William Faulkner – from Oxford Mississippi to the world of the studios in 1942 - or Truman Capote or F. Scott Fitzgerald, let alone Thomas Mann) and Topanga Canyon was a world apart from near by bioinvasive rich palm-laden Beverly Hills. Humphry Repton might well have recognized the rectilinear approach to alphabetized streets and well-manicured lawns and sidewalk bordered palms that adorned much of West Los Angles, but he would have preferred Lofting's house up in the mountains of Topanga. However, American publishers would not accept Lofting's not-for-children publications, like "Victory for the Slain" and – given England's predicament during the War, it is incredible that the English Press were so open to it.

Of course, by World War II, Doctor Dolittle was one of the most instantly recognized names in the world, vying with Churchill and Gandhi for notoriety. And it was precisely Churchill's vigorous support of the War that Lofting took issue with in his poetic outpouring. Not unlike George Orwell whose stinging critique of Gandhi's naivety regarding Hitler rang perfectly true but went largely unnoticed, Lofting's anti-war approach has enjoyed scant comment by critics, but such views were totally aligned with the character of Doctor Dolittle.

There is little doubt that Lofting had a significant impact on Orwell, and, quite possibly, on Gandhi himself. Lofting had re-married Canadian Josephine Fricker and moved to Los Angeles in 1935. In 1936, the author's third child was born, the only one to he and Fricker, Christopher Clement. Olga Michael Fricker, Josephine's sister, would become Lofting's secretary, and later on, the posthumous editor of his manuscripts.

"A Visit Back in Time"

We visited Lofting's Topanga Canyon abode, looking out over the Pacific five miles or so away, as the Condor flies. There were approximately 3,000 people living in Topanga Canyon when Lofting moved there. The great author had a studio separated from the house and we discovered rumors of a lost manuscript that he had hidden under the fireplace mantle, though his son Christopher who spoke to us in his office in Manhattan late one afternoon had had heard nothing about such a work.

Biographical and critical studies of Lofting to-date are extraordinarily meager. The one substantive work is by the aforementioned Gary D. Schmidt, professor of English at Calvin College in Michigan. [*17] An additional but paltry array of materials available to Schmidt underscores the strange case of a somewhat "missing" life. Seven pages of material published in an obscure book a year after Lofting died, by Helen Dean Fish; seven additional pages in the little known magazine, "Elementary English," in April, 1955, a page-and-a-half by Colin Lofting, the son, in a September 30th, 1966 issue of Life Magazine, and finally, a monograph, published in London by Edward Blishen in 1968 - a pamphlet so obscure as to be found today in the odd rare book rooms of a few universities. In addition, there are at least 10 critical studies of children's literature in general that refer to Dolittle, one even suggesting, according to Schmidt, that **The Story of Doctor Dolittle** was flawed because of "scientific inaccuracies." [*18]

Even within Topanga Canyon, in greater Los Angeles, where Lofting spent the last twelve years or so of his life, the local Historical Society did not exactly seek importunity for the elusive giant's whereabouts though one weekday afternoon, we were kindly shown the Lofting family's rugged 9-acre estate and without too much difficulty were able to piece together what happened to the Doctor who sold it to Lofting - a famed archaeologist known for digging up mummies. We wandered around the Lofting acreage, looking out over the Pacific, in awe of the extraordinary presence that still seemed to rest comfortably there in the name of a man whose personal world has languished in so much obscurity.

"2,500 Pages for Posterity"

But what is most clear about the long-lasting appeal of the Dolittle books – nearly 2,500 pages for posterity - and evident character behind them - is this uncanny, unambiguous legacy of love of nature. The character of Dolittle himself was not consistently vegetarian yet the founding member, by words and actions, of the animal liberation movement. He thus set a profound example to the rest of humanity about what steps individuals can take to make this a better world. His impulses were neither flagrant nor devout, but stemmed from an abiding moral outrage that translated into household particulars, and the ability to set sail to Africa - or to the moon - without so much as the hint of a doubt as to any known outcome. A perpetual Debussy "Girl With The Flaxen Hair" in ease, elegance and harmony.

Dolittle did not distinguish between the "cat-meats man" and a parrot; gave equal time to a duck or a mule; had little care of money or housekeeping, yet, by all accounts was focused on worldly things. If he was a metaphysician, steeped in history, logic, veterinarian medicine, Leicester Square, art and science, he did not wear such learning on his sleeves. This made his books bestsellers.

T.S. Eliott once argued that nothing is original. That all art derives from other art. If this is true, then it is equally critical to recognize that Lofting takes his place in a moment and a time when the reading culture, and children - as well as adults - everywhere were ripe for the Dolittle universe. This did not just happen because of the horrors of World War I and the grip of the massive influenza pandemic, which combined would leave at least twenty-five million people dead (and who knows how many of their companion animals); dead and homeless. In spite of those sorrows, or because of them, the world had always harbored a deep and enduring love of nature, of other species, of the idea of paradise; and Doctor Dolittle allowed us to inhabit that world of adventure and compassion without risk, or uncertainty. [*19]

He died September 26th, 1947, in Topanga Canyon. A mere 61 years old. He is buried at the Evergreen Cemetery in Killingworth, Middlesex County, Connecticut. The genius that was Hugh Lofting will never leave us.

Footnotes:

*1 http://www.brainpickings.org/index.php/2013/09/16/e-b-white-dog-obituary/

*2 *See p.196, **God's Country: The New Zealand Factor**, by Michael Charles Tobias and Jane Gray Morrison, A Dancing Star Foundation Book, Zorba Press, Ithaca New York, 2011.

*3 http://andalusiafarm.org/andalusia/environmental_edu.htm

*4 http://www.georgiaencyclopedia.org/articles/geography-environment/william-bartram-georgia

*5 For other references to this gorgeous bird, see the many Derbyian links compiled at http://harleysperch.com/darbypics.htm.

*6 "Your lordship's most obliged servant": letters from Louis Fraser to the thirteenth Earl of Derby, 1840 to 1851," by Amberley Moore, Editor of the Ibis, http://www.euppublishing.com/doi/abs/10.3366/anh.2004.31.1.102)

*7 See the 2001 publication of **Threatened Birds of Asia –The BirdLife International Red Data Book**, edited by N.J. Collar, A. V. Andreev, S. Chan, M. J. Crosby, S. Subramanya and J. A. Tobias, BirdLife International Publishers, London. See also, "Derbyan Parakeet (Psittacula derbiana): information on the threat from trade and resulting population trends requested," Posted on December 6, 2010 by Andy Symes, http://www.birdlife.org/globally-threatened-bird-forums/2010/12/derbyan-parakeet-psittacula-derbiana-information-on-the-threat-from-trade-and-resulting-population-trends-requested/.

*8 See "IUCN Red List of Threatened Species" http://www.iucnredlist.org.

*9 London: J. M. Dent & Sons 1920.

*10 See P.L. Sclater, "On the general geographical distribution of the members of the class Aves," J. Proc. Linnean Soc. Zool. 2 (1858) 130–145.

*11 See "Osbert Salvin [1835-1898] naturalist" by Clemency Thorne Fisher in Dictionary of National Biography, online. See also, the feature documentary by Michael Charles Tobias and Jane Gray Morrison, Hosted by Russell Mittermeier, "Hotspots," a Dancing Star Foundation Film, KQED/public broadcasting, 2008: www.hotspots-thefilm.org.

*12 See **Curiosities Of Ornithology**, by T, W. Wood, London: Groombridge And Sons, 19th century, n.d., p.14. See also, Tim Birkhead's **Bird Sense – What It's Like to Be a Bird**, for a fascinating discussion of filoplumes. Walker & Company, New York, 2012, pp. 84-86. Also, in Birkhead, p.24, for a discussion of just what colors within feathers birds might well see, that we do not.

*13 Translated by A. Teixeira De Mattos, Illustrated by E. J. Detmold, London: George Allen & Co. Ltd., 1912.

*14 p. 66, **Doctor Dolittle's Garden – Told & Pictured By Hugh Lofting**, Published by F. A. Stokes Co., New York, 1927.

*15 ibid., p. 76..

*16 For all of this data see, http://www.lamountains.com/parks_plants.asp

*17 Hugh Lofting (Twayne Publishers, Maxwell Macmillan, Toronto, Canada, 1992.

*18 E. Cameron, **The Green and Burning Tree**, Boston: Little, Brown, 1969, pp.49-56.

*19 See **The Story of Dr. Dolittle** (1920); **The Voyages of Dr. Dolittle** (1922); aka **Doctor Dolittle and the Pirates; Doctor Dolittle's Post Office** (1923); **Doctor Dolittle's Circus** (1924); **Doctor Dolittle's Zoo** (1925); **Doctor Dolittle's Caravan** (1926); **Doctor Dolittle's Garden** (1927); **Doctor Dolittle in the Moon** (1928); **Gub Gub's Book** (1932); **Doctor Dolittle's Return** (1933); **Doctor Dolittle's Birthday Book** (1936); **Dr. Dolittle and the Secret Lake** (1948); **Doctor Dolittle and the Green Canary** (1950); **Dr. Dolittle's Puddlesby Adventure** (1952);

Noisy Nora: An Almost True Story (1929); **Porridge Poetry: Cooked, Ornamented, and Served Up** (1924); **Tommy, Tilly, and Mrs. Tubbs** (1936); **Victory for the Slain** (1942); **The Twilight of Magic** (1930); **Dr. Dolittle's Animals** (1998); **Dr. Dolittle to the Rescue** (1999); **Doctor Dolittle and the Lighthouse** (2000); **Doctor Dolittle in Trouble** (2000); **Doctor Dolittle Saves the Day** (2000)**;** **Doctor Dolittle Takes Charge** (2000); **Doctor Dolittle's Ambulance** (2000)**;** **Doctor Dolittle's First Adventure** (2000).

The Authors recommend that readers also confer with the following, to better grasp the evolution of ethology leading to Hugh Lofting: Gottfried Immanuel Wenzel, **Neue auf Vernunft und Erfahrung gegrundete Entdeckungen uber die Sprache der Thiere**, Wien, Anton Doll, 1800; **A Philosophical Amusement Upon The Language Of Beasts**, by Father Guillaume-Hyacinthe Bougeant, (London: Printed for T. Cooper, 1739); Joseph Ritson, **An Essay on Abstinence From Animal Food, As a Moral Duty**, London, Printed for Richard Phillips, St. Paul's Church-Yard, 1802; **The Speech of Lord Erskine in the House of Peers on the Second Reading of the Bill for Preventing Malicious and Wanton Cruelty To Animals, Taken in Short Hand**, Edinburgh, (Printed for Alex, Lawrie, Bookseller and Stationer, Parliament Square, 1809); Percy Bysshe Shelley, **Queen Mab,** (Printed and Published by W. Clark, London,1821); Thomas Taylor, **Select Works of Porphyry: containing his Four Books on Abstinence from Animal Food; his Treatise on the Homeric Cave of the Nymphs; and his Auxiliaries to the Perception of Intelligible Natures, and an Appendix, explaining the Allegory of the Wanderings of Ulysses,** (London, Thomas Rodd, 1823); W.D. Fellowes, Esq., **A Visit to the Monastery of La Trappe**, in 1817 (London, 1818 (2nd. Edition); See also www.humane-history.org.

4

A MYSTIQUE OF WATERFALLS

"First Revelations"

As indicated in Shinto tradition, where a waterfall can easily be described as a living spirit, believed in, with that divinity, or Kami, fully in mind, it is incumbent upon the imagination to find reconciliatory approaches to reverence that can expand the aegis of compassion so as to recognize the entire ecosystem involved in any deification. To that extent we have seen, for example, His All Holiness Bartholomew I, leader of the world's approximately 300 million Orthodox Christians, declare that harming other species, degrading the natural world, is a "Sin" (the 1997 Santa Barbara Declaration). Similarly, in early July 2014, whilst speaking at Italy's University of Molise, Pope Francis called it "our sin" for wreaking havoc with the Earth's environment, putting all people on notice and calling it the way forward "one of the greatest challenges of our time."

Just as the many spiritual revelations of every known ethical tradition have paid diverse tribute to the environment, our celebration of beauty herself may be key to our species' survival. The artistic reveries of a Leonardo, who first indoctrinated landscape paintings with aerial perspectives over mountain ranges and glaciers –setting the stage for a LANDSAT and Google Earth - and focused with equally enormous detail and energy on the backgrounds – whether in his 1474 painting of the Florentine "Ginevra de' Benci" or his "Virgin of the Rocks" (1483-1486), or "Mona Lisa" (1503-1506) - he enlisted beauty on behalf of scientific enquiry and natural history, encompassing in his realm of vision whole ecosystems. This aesthetic inculcation of a greater schema represents a milestone in the history of conservation prospects. Putting a person on the moon and looking back at Earth would have been a familiar sight, of sorts, to Leonardo.

Human perspectives are not unique, though certain individuals clearly see beyond. The same is true among members of all species. Bill Wallauer, a videographer working for years filming chimpanzees at Gombé, in Tanzania, often observed the chimps around a waterfall, one chimpanzee in particular named Freud. Following a complex display, Freud crossed repeatedly over the falls, rolled stones down the cascade wall, descended on vines overhanging the falls, and then settled into a prolonged gaze over to the cataracts, as if mesmerized, or deeply contemplating. Writes Wallauer, "He (Freud) relaxed, then turned to the falls and stared at it for many minutes." And Wallauer adds, "...the displays in reaction to these elements of nature suggest that chimpanzees find something meaningful which could possibly be described as reverence to aspects of their environment." [*1] Wallauer writes that he craved to know what was going on in Freud's mind.

Every individual, regardless of the primate or other species, is likely to have a different response. When anthropologist Michael J. Harner studied the Jivaro of the Ecuadorian Amazon, and followed villagers to nearby waterfalls, he discovered that for a young male, initiation in the waterfalls was key to that youth's survival as a Jivaro. Only at the waterfalls would a spirit vision come to the youth, that spirit, or soul, and its accompanying spirit encounters (Arutam wakani) the source of a Jivaro's longevity and immunity to death. One soul visitation from the waterfalls meant that the Jivaro could never be killed by a fall, or a blunt object to his head. Two souls guaranteed immortality. [*2]

A mist-enraptured spectacle of catalytic conceptualizations hovers about, within, and across the pages of waterfall discovery and meditation. There can be no more elusive yet permanent reminder of our addiction to this mystery that is nature than in the ever-enlivening unresolve of a waterfall, one of nature's grandest paens to herself. How explorers and artists, philosophers and scientists, ecologists, mystics and indigenous peoples have worshipped and thrilled to the surreal promise of waterfalls in their lives re-enforces the instinct that is embryologically protective, eliciting that cohabitation of the psyche with the Earth.

This connection is undeniable. Having experienced the waterfall, we want to safeguard every tangible and intangible nuance of which it is comprised: the declivity, the thunder, its height, breadth, spray, effluvia, itinerary, bottom-lands, the pools it may form, the creatures hidden beneath the boulders, all that dwells, flies, feeds, cohabits its catacombs of falling waters.

From Frank Lloyd Wright's famed Fallingwater House near Pittsburgh, in southwest Pennsylvania's Laurel Highlands, to all those thousands of modest tumbling creeks in states like Tennessee, the sensation that drives the mystery itself is one of pure philosophy acting out its self-accord. The waterfall is like a chime in the wind that self-propels its mission without requiring additional affirmation. This is a self-contained Imperative that inheres every natural phenomenon associated with dihydrogen oxide and gravity, conveying itself without need of elaboration to the onlooker. By dint of this indisputable confirmation in thought and joy, the heart becomes embedded. The fervor of this emblem strikes of a unique loyalty that may be the essence of metaphysics. Waterfalls are also among the most difficult of natural phenomena to protect. Their every second defines elusiveness.

There are, of course, less heady predilections arising from the concourse of a thundering body of water. No love affair with the cliff, no insight that smacks of the beyond. A few have set forth little more than a populist, anti-establishment dare-devil's orientation, as in the early 19th century case of Sam Patch "the famous jumper" as historian Paul E. Johnson writes of him. Patch survived his first leap off Niagara Falls before a multitude of incredulous spectators. His second jump from the same falls was not nearly as successful. He died, joining some 5,000 others who have perished in waters that (these days) move at approximately 150,000 gallons per second. [*3] There have been tragic video clips posted of various suicides, from bridges, and from Niagara Falls. [*4] Notwithstanding the unseen visceral horror, or imponderable impact on surviving loved ones, beyond the fascination humanity seems inclined towards in viewing disasters, there is the strange attractor that is human contact with a waterfall. It cannot be denied.

The one person who has survived a jump off Niagara, without the aid of a barrel or any other protective gear, was a gentleman named Kirk Jones. On an evening in October 2003, Jones set himself upon a headlong course off the 187-foot high cliff of thundering falls. On his return to dry ground down below, he was arrested, placed under psychiatric evaluation for three days, fined $5,000 Canadian dollars, and banished for life from the Canadian side of the Falls. He also obtained a $100,000 contract to join a circus. [*5]

Others have inadvertently lost everything to waterfalls, and the death-song of the famed Wenona, as given in Dacota, or "Legends of the Sioux," at the Falls of St. Anthony in Minneapolis are a case in point; how the loving wife, while in her canoe, was accidentally swept over the falls which seemed to entrance her such that even the cries of her husband on the adjoining shore failed to "rouse her from her fatal purpose." [*6]

Still others sought waterfalls for one unambiguous goal: to merge their lives with a hopefully transcendental ending in the action of suicide.

There have been countless leaps, by lovers or otherwise, off waterfalls in Yosemite, at Niagara, and at a remote site in India's state of Andhra Pradesh, where now a person stays devotedly on guard beneath the falls, much like surveillance cameras on the Golden Gate Bridge. M. Bheemanna, the watchman in question, has saved at least a dozen lives at the Pochera waterfall. [*7]

"Ecological Threats"

It might seem to enter the Pure Realm of meditation and philosophy when we ask: How does one protect the flow of water? But this is no Zen Koan. Countless real-world waterfall-related dangers and ambiguities lurk, in the form of human engineered riparian diversions and dam building. Hydrologic intervention is not merely about the greed for hydro-electric power, or short-sightedness of many of the projects (half the dams in India, for example, now obsolete). New dams recently constructed, for example, in California, await rain water which – in the worst drought in centuries – appears nowhere in sight. But what is actually at stake in any dam-related venue is the conservation and biodiversity surrounding waterfalls, which translates into countless lives of other species, most of whom remain undocumented or researched.

Since the 1950 Niagara River Water Diversion Treaty between the U.S. and Canada, roughly half of all water that had previously flowed over the falls – those perturbatious and enlightening chimeras painted by George Inness and Frederic Church, among countless others, and making Niagara one of the great spectacles in the world, has been lost, for purposes of hydroelectricity distributed between the two nations. If Niagara is still awesome, it is so despite the zealous efforts of humanity to diminish everything about her. What is not obvious to most visitors, however, is the "loss of fish and wildlife habitat throughout the Niagara River Area of Concern (AOC)," and this has been cited as "dramatic". Because the Niagara River Corridor has been designated a Global Level Important Bird Area (supporting up to "100,000 gulls of up to 10 species, tens of thousands of waterfowl up to 34 species, and an incredible diversity of neotropical migratory songbirds") not to mention fish, reptiles and amphibian species that are endemic to the Niagara, human impacts have been horrendous. [*8]

Elsewhere, other dramatic changes have been enacted by governments that felt they had better uses for waterfalls than nature intended, from what was once one of Australia's greatest waterfalls in the Barron Gorge near Cairns, transformed to a mere trickle; at the Maria Cristina Falls near the city of Iligan on the Philippine island of Mindanao; at Sweden's Trollhattan Falls, and elsewhere. Jamaica's famed Dunn's River Falls were spared destruction by the energy-hungry government, while many other Jamaican falls were not. But now, with over one million visitors per year to the Dunn's River site, and many people attempting to climb the falls, one landscape architect has described the situation as "a tragedy in three acts". [*9]

At Tanygrisiau Falls, 190 miles northwest of London, a clever compromise measure involving the building of two reservoirs managed to save the water, at the same flow levels that would otherwise have been sacrificed from the cliff to the grinding mechanisms of power turbines. Inga (Livingstone) Falls, one of 32 cataracts along the all but non-navigable lower Congo River [*10] have already been partly transformed, with Democratic Republic of the Congo blueprints in place to make what is the largest flow of water on earth (1,500,000 cubic feet per second) into the biggest hydroelectric power project anywhere.

Nearby, in Uganda, plans have been in the works for many years to dam the Bujagali, just under Lake Victoria on the Nile. Claims by the National Association of Professional Environmentalists, a Ugandan NGO, that the project would be disastrous have slowed down, but by no means impeded the Ugandan government's intent to drown the (sacred) Bujagali Falls, a national treasure with huge biodiversity, in order to reap projected hydroelectric gains, with promised support from the World Bank and others. [*11]

In 290 BC the Romans actually created a waterfall for their pleasure, by diverting the Velino River over the Marmore Cliff. Then, in 1950 AD, the Italian government furthered this interventionist zeal by adopting a utilitarian approach that saw greater benefit from energy than nearly 1,700 years of beauty.

The construction of more modest, less ecologically-challenging waterfalls became an art form in Japan, as described in the first Japanese work on garden design, the **Sakuteiki** by Tachibana Toshitsuna (1028-94). Some of the earliest gardens, like those of Akasaka Rikiu, and Koraku En, in Koishikawa, and the cascades of Mangwanji in Nikko, are as philosophically arranged for natural and aesthetic consequences as were the five small waterfalls designed for Central Park by Frederick Olmsted and Calvert Vaux. Miniature waterfalls are one of the key design tools for landscape architects throughout the world, the aforementioned Frank Lloyd Wright's Fallingwater House being among the most celebrated and, in Wright's case, involving an in situ cascade. Most recently New York City's East River sprouted four temporary waterfalls, designed by Olafur Eliasson, and each between 90 and 120 feet in height, a nod to the modern sublime in the form of radical public space. [*12]

But the loss, or diversion of waterfalls at all scales – whether in the Congo, amid the scores of artificial dams of concrete erected by the Tennessee Valley Authority, or during the tragic loss of six small waterfalls above Altadena, in Southern California's Rubio Canyon in 1999 - can have severe impacts on biodiversity, as with the destruction of any sensitive habitat involving water. [*13]

The relationship between a fresh water vertebrate (a fish) and a dam made marked international news when, on August 12th, 1973, a rare snail darter (*Percina tanasi*) was discovered by zoologist David Etnier who was snorkeling in the Little Tennessee River, in research pertaining to the construction of the Tellico Dam. His findings, and the resulting political firestorm – a dam versus a little fish - would go to the United States Supreme Court and form the basis for the most extensive, heated and melodramatic legal case in American environmental history. [*14]

For all of our celebration of waterfalls in art, as cornerstones of national parks, such as in Yosemite and Yellowstone and New Zealand's Fiordland National Park, and despite a huge history of systematic exploration of the geographical origins and seasonal variations of waterfalls, the ecology of these phenomena remain enigmatic, whether on Earth or elsewhere, even on Mars, where the frozen 12,000 foot deep Echus Chasma suggests there were once waterfalls of outrageous magnitude; not to mention other NASA images of frozen Martian mountain ranges in the form of vertical gullies hundreds, possibly thousands of miles high that might have given vent to unimaginable cascades. [*15]

"Biological Habitats"

Waterfall biodiversity involves human settlement as well as a staggering plethora of symbiotic, water-dependent organisms, as the aforementioned biological riches of Niagara suggests. To take one of the oldest known waterfall habitations, among humans, consider Victoria Falls on the Zambezi River between Zambia and Zimbabwe. Our ancestors have been living there in the mist and rainbows of Victoria for at least 3 million years. There is ample documentation supporting the conclusion that one culture after another made Victoria Falls their home, continuing through the Middle and Late Stone Ages, up until today.

But the human presence is but one species, amid an onslaught of others at such places, and new ones being discovered all the time. Recently, for example, a never-before-witnessed minute aquatic beetle family, the Meruidae, also known as "comb-clawed cascade beetles" was discovered by two U.S. scientists along the Rio Coromoto in the Guyana Shield of Venezuela. [*16]

In July of 1972, a biologist with the University of Tennessee at Chattanooga discovered one of the greatest known stands of white pines adjoining Piney Falls in a deep gorge of the Cumberland Plateau in Rhea County. In the West Tropics Bioregion of Australia, the Waterfall Frog (*Litoria nannotis*) is restricted, as its name implies, to exposed stream positions. Listed as Endangered under that country's Environment Protection and Biodiversity Conservation Act 1999, the frog's alarming decline appears to be related more to amphibian chytrid fungus than habitat loss, given that this latter threat to a large-scale ecosystem has been largely rectified. [*17]

In the case of the remarkable waterfall-laden gorges of Bioko, Equatorial Guinea's Atlantic jewel of an island, many of the plunging, heavily vegetated rivers remain unexplored, except by helicopter. [*18] The same case holds for much of Bhutan, southwestern China, southwestern Chile, large parts of Canada, Alaska, Norway, Greenland and the Antarctic.

But even at all the great waterfalls of the world that are well documented, measured in every direction, and visited by mobs of thirsting adventurers, the other life forms present are often obscured or entirely unknown. This is especially the case in tropical and temperate falls, like Victoria, the remote Tres Hermanas in Peru (3,000 feet), Olo-upena Falls in Hawaii (2,953 ft) and especially the much-celebrated Angel Falls in Venezuela (which, at 3,212 feet of vertical drop makes it the tallest waterfall in the world). Angel (Kerepakupai merú) in Canaima National Park, Bolivar State, is cloaked in mostly endemic vegetation. The wall behind Angel (the Auyantepui, that has been climbed) and the falls plunging over it, have long been sacred to the Pemon peoples. But the outside world only learned of them when the American aviator James Crawford Angel flew over the mist-encased phenomenon in 1933 in search of precious minerals. In 1937 Angel, accompanied by his wife Marie and three friends actually crash landed on the summit and spent 11 days trekking out. Why the waterfall was named after Angel must conjure mixed feelings given what we know today to be a highly vulnerable ecosystem. For example, along the rim of the Pakaraima Mountains throughout the Venezuela/Brazil border are countless waterfalls, including the second highest in the world, Kaieteur, at the intersection of the enormous Roraima cliffs and tropical lowlands. This is where Sir Arthur Conan Doyle imagined his Lost World. But it is also home to a remarkable proliferation of sensitive bromeliads, tiny terrestrial bladderworts, herbs, lichens, and countless tree species, with tree and epiphyte families "poorly represented" in terms of documentation to date. [*19]

From southern Georgia and little-known parts of Virginia and Indiana, to Ecuador and Japan, people are not the only ones who flock to waterfalls. Birds, insects, reptiles, spiders, fish, crustaceans, and other vertebrate all have made waterfalls their unique home for millions of years. The chances of discerning new species and ecological relationships in all these waterfalls is undeniable. Moreover, new cascades of surprisingly vivid scale, yet heretofore undocumented, continue to be discovered, including a not insignificant one in California within the past decade.

The greater question that must attend upon every expedition, as well as much eco-tourism, are guidelines for inhibiting possible damage to these exquisite environments. Just as the French have all but closed off the great painted caves, such as Lascaux, and the Aurignacian Chauvet [*20] and Native Americans and other indigenous peoples throughout the world have rendered their sacred petroglyphs inviolate to traffic – as in the case of the Chumash Painted Rock and Heritage sites above Santa Barbara (the so-called "Painted Cave"), or on the Carrizo Plain National Monument in Central California - waterfalls should be duly re-considered in every instance by those charged with their care. At the very least, human visitations to waterfalls must be appropriately considered in no less manner than our journeys into churches, across wetlands, into arboretums, or clean rooms.

"Continuing Expeditions"

Long before remote sensing would reduce some of the risks and probably much of the exhilaration of old-time exploration, people like British botanist Francis Kingdon Ward (1885-1958) would spend nearly a year exploring the Tsangpo (Brahmaputra) Gorge between the (then) North East Frontier Agency of India and Tibet, in search of rare plants and the rumored highest waterfall in the world. [*21]

Many of the "KW" Himalayan and Chinese rhododendron species would be transplanted in gardens like Kew, in London. But his most telling legacy is his failed search for that legendary waterfall.

Others, such as Sir Henry Morton Stanley (1841-1904), who was played by Spencer Tracy in the famed 1939 epic movie "Stanley and Livingstone" would be the first to descend the impossible Congo, just as Livingstone would be the first European to lay eyes upon Victoria Falls. Stanley's harrowing 999 days on the Congo saw him reach the mouth of the horrendous river on August 9, 1877, after losing roughly 2/3rds of his expedition, 242 people, to the endless series of man-eating waterfalls, as they were enthusiastically called. [*22] Unlike Ward, who leaves a record of ecological integrity, Stanley's sad aftermath continues to be burdened by posterity's uncertainty with respect to his naïve complicity in King Leopold II's genocidal empire building in Central Africa.

These two "styles" of exploration – small, intimate, scientific movement through the wilds, versus huge expeditions exacting heavy impact – are symptomatic of the challenges facing new waterfall exploration, as individuals continue to search for the next highest, widest, most dangerous, beautiful and bountiful waterfalls in the world. Climate change will certainly exert freakish and profound tweaks in the hydrological patterns across the planet, and those areas with many of the most tortuous altitudinal differentiations, such as the Himalayas and Alaska Range, are likely to see some of the most extraordinary perturbations, with potentially calamitous impacts on waterfall biodiversity. Alaska is already heating up at least double the speed of the lower 48 states, whilst reflectivity on Greenland's ice surface is having a 4-to-5 times rate of ice melt, compared with areas farther to the South. This state of affairs places an unusually critical burden on ecologists who would aspire to research unique falling water habitats.

As climate change accelerates even the pH of the water will be altered, species thrown into more and more flux. With changing precipitation patterns and accelerated snowmelt, new artistic nuances will undoubtedly emerge, as well. Historically, waterfalls have altered their own rock walls and chasms, such that form and function – when the action of continuously pounding water on vulnerable geologic formations is involved - have always been engaged in a veritable dance. The Colorado River wending through a billion years+ of Grand Canyon geological history is one noteworthy example. Other water-formed gorges may be as diminutive as the cascades portrayed in a small but stunning "Rydale Falls," photographed in England in 1870, [*23] but all are vulnerable, regardless of size.

"The Art of Protecting Water"

Tens-of-thousands of photographers, painters, and poets have waxed, made pilgrimage, and focused upon waterfalls. 19th century John Gibson wrote perhaps the first glowing account of the great waterfall explorations [*24] while inveterate champion of cascades, Edward C. Rashleigh, lost his footing and was swept down the Orange River towards the rim of the massive 183-foot Aughrabies Falls of South Africa, an account written up in local newspapers at the time. [*25]

The late 17th century Dutchman, Johanne Herbinio, composed the first western treatise on waterfalls, hydrological anomalies, and the very waterfalls in Paradise, providing actual maps to their whereabouts. [*26] Herbinio's contemporary, the painter Jacob van Ruisdael, transformed the Netherlands into a kind of Japanese Otherworld with his own torrents, much in the manner of a Hokusai, who painted the waterfalls of Japan. British spectacle painter, John Martin, who, in turn, had a marked influence on the poet Percy Shelley, glorified an incandescent imaginary alpine cascade in his great painting "Zadak In Search of the Waters of Oblivion" (1812). Interestingly, John Martin – who painted mountains that have been calculated to be approximately 15,000 feet in height – actually never left metropolitan London. His extraordinary Romantic wilderness was utterly invented. His heart was in it.

Even Sir Arthur Conan Doyle succumbed to the obvious power of plunging cascades, not just in his South American romance of surviving dinosaurs atop a waterfall-laden cliff, but when he chose a real location near the town of Meiringen, in the Bern Canton of Switzerland, the Reichenbach Falls, as the site for the final conflict and (initially alleged) death of Sherlock Holmes, in his short story, "The Adventure of the Final Problem" (1891). Doyle wrote of Reichenbach (a long time haunt of enthusiastic tourists) as forming "an immense chasm, lined by glistening coal-black rock, and narrowing into a creaming, boiling pit of incalculable depth, which brims over and shoots the stream onward over its jagged lip." Holmes' physical hand-to-hand combat with Professor Moriarty resulted in Holmes' plunge from off a ledge on the side of the 100-meter falls. Twelve years later, fanatical fans persuaded Doyle to bring back his sleuth. Doyle conceded by suggesting that his hero had actually survived that fateful day years before, by grabbing onto some vegetation during his plunge and thus saving himself in secret. [*27]

No matter what secrets, biological tumult and aesthetic alterations lie in store for the planet, waterfalls will remain a brilliant source of pilgrimage, fascination, and, perhaps most importantly, a tonic to be safeguarded, as they were for Ottorino Respighi (1879-1936) in his musical cantata, "La Primavera," and the great Chinese painters of the "shan shui" style. That particular juxtaposition of elemental figurative as well as abstract forms, emerged during the early 11th century Sung Dynasty and came to herald some of the greatest works of art in Chinese history. Kuo Hsi's masterpiece, the hanging scroll entitled "Early Spring" (1072) in the National Palace Museum in Taiwan is representative of this Chinese renaissance. Within the enormous glowing canvas are a series of tenuous, fragile pathways, severe overhanging cliffs, bare uprooted trees clinging to rock and, at the heart of the massive composition, a bursting waterfall. [28] This symbol for the renewal of life which has come to dominate Asian aesthetics – and all protection in any guise - may also prove to be the ultimate meaning of waterfalls for all those who behold them, humans, Alaskan grizzly bears, snail darters, water beetles, and John Muir's water ouzels. [*29] They are, like a lamb in a great Van Eyck, the harbingers of iconic centrality across our whole vision of nature.

Footnotes:

*1 www.janegoodall.org/chimp_central/chimpanzees/behavior/rain_dance.asp.

*2 See **The Jivaro – People of the Sacred Waterfalls**, Doubleday, Natural History Press, Garden City, New York, 1972, pp.134-136.

*3 See **Sam Patch: the Famous Jumper**, by Paul Johnson, Hill & Wang Publishers, New York, 2004.

*4 www.youtube.com/watch?v=GdPxrVwtaes

*5 See journalist Shawn Windsor's story in the Detroit Free Press, "Kirk Jones –the only person ever to survive an unprotected trip over Niagara," http://www.bookofjoe.com/2004/10/shawn_windsor_o.html.

*6 **The Romance Of Indian Life**, by Mrs. Mary H. Eastman, Lippincott, Grambo & Co., Philadelphia, 1853, p. 246.

*7 See "The guardian angel at Pochera," by S. Harpal Singh, www.hindu.com/2006/12/stories/2006061202520200.htm.

*8 See "Problem Statement" in "Little Beaver Is. Wetland Restoration: Ecological Modeling," GLNPO ID: GL2000-012, USACE-Buffalo District, New York.

*9 "Fall of beauty: The story of a Jamaican waterfall –a tragedy in three acts," by Brian J. Hudson, Tourism Geographies, Volume 1, Issue 3, August 1999, pp.343-357.

*10 See www.britannica.com/EBchecked/topic/288028/Inga-Falls

*11 See "Controversy surrounding Bujagali dam persists as Lake Victoria…" www.bicusa.org/en/Article.3404.aspx. A similar problem - of huge historic consequences – looms over the Bekma Dam Project in northern Kurdistan in Iraq –referenced in an earlier chapter - the most important of the Shanidar caves in a direct beeline with the long-time dam construction. See http//www.flickr.com/photos/one-thirteen/sets/72157622721849593/

*12 www.nycwaterfalls.org/#/Olafur_Eliasson/Interview_With_Olafur

*13 http://www.aaaim.com/echo/v4n2/index.htm

*14 See "Tellico Dam still generating debate," www.knoxnews.com/news/2008/Apr/13/tellico-dam-still-generating-debate/

*15 See http://www.marsanomalyresearch.com/evidence-reports/2003/050/50-waterfalls-report.htm.

*16 See "A new aquatic beetle family, Meruidae, from Venezuela [Coleoptera: Adephaga]," by Paul J. Spangler and Warren E. Steiner Jr., Systematic Entomology (2005), DOI:10.111/j.1365-3113.

*17 See www.environment.gov.au/cgi-bin/sprat/public/publicspecies.pl?taxon_id=1817.

*18 See "YouTube- Bioko Island Equatorial Guinea," www.youtube.com/watch?v=BhW7tiq5r1l.

*19 See "Introduction to the Kaieteur Falls National Park Plant Checklist," www.mnh.si.edu/biodiversity/bdg/kaieteur/intro1.html.

*20 See "First Impressions -Letter From Southern France," Judith Thurman, The New Yorker, June 23, 2008, pp.58-67.

*21 See Ward's **The Riddle Of The Tsangpo Gorges**, Edward Arnold, London, 1926.

*22 See Stanley's **Through The Dark Continent, Or, The Sources Of The Nile Around the Great Lakes of Equatorial Africa**, Harper & Brothers, New York, 1878.

*23 **Our English Lakes, mountains, And Waterfalls, As Seen By William Wordsworth, With Photographic Illustrations by Thomas Ogle**, 4th edition, Provost & Co., London, 1870, p.50.

*24 **Great Waterfalls, Cataracts and Geysers**, by John Gibson, T. Nelson And Sons, London, 1887.

*25 **Among The Waterfalls of the World**, by Edward C. Rashleigh, Jarrolds Publishers, London, 1935.

*26 **Dissertationes De Admirandis Mundi Cataractis Paradiso**, by M. Johanne Herbinio, Janssonio-Waesbergios, Amsterda, m, 1678.

*27 http://sherlock-holmes.classic-literature.co.uk/the-final-problem/

*28 "National Palace Museum," www.npm.gov.tw/en/collection/selections_02.htm?docno=48&catno=15.

*29 **The Yosemite**, by John Muir, The Century Co., New York, 1912, p.22; ibid., p.29; See also **Yosemite Legends**, by Bertha H. Smith, With Drawings By Florence Lundborg, Paul Elder And Company, San Francisco, 1904, p.43. See **A Landscape Book, by American Artists and American Authors**, G.P.Putnam & Son, New York, 1868, pp.29-31.

5

EXISTENTIAL BIOLOGY

"The Ecology of Introspection"

Our investigation of natural history, from the earliest human evidence for observation of other species, landscapes and climatic events, has tended to express its shamanic or empirical findings by way of signaling to others some precise message. Codes, ambiguities, subtleties: these are today's interpretive language for what may well have been a degree of precision unclouded by motive or confusing context. The change of seasons. The migration of large animals, and the habits of small ones. Unique landforms. Galaxies and comets. The direction to potable water, and the shimmering beings of another kind therein. Auras, transfigurations, visions. These semiotic emblems of our hyper-interface with everything around us, had been, until just a few millennia ago, seemingly quite in balance between an inner openness and outward willingness to fall into tune with the harmonies of the natural world. Our personal aspirations were not in conflict, as a rule, directly or indirectly, with ecosystems.

The resulting human creations and entablatures have always struck a large contingent of "modernists" as somehow credulous, or "primitive" (the word of choice still in currency amongst professionals in fields as diverse as ethnography, engineering, astronomy, anthropology, the arts and the health sciences).

But the only truly "primitive" displays by humankind have emerged in the wake of disproportionate power it has wielded – *we* have wielded -over others, as performed by one genocide or murder after another, served up in such a cacophony of guises as to defy any testable quantification, certainly by several orders of magnitude.

Whereas our wilder forebears – and some remaining contemporaries – have naturally come by this outward and multi-various design in life; a confluence of inner soul and outer satisfaction by no other reliance than that upon which nature in her bountiful manner has forever given to us, asking for only the basics of respect and moderation. An organic gardener understands this well. But this basking in what has been given since birth is not the modus operandi of a species heading rapidly towards ten billion or more high-end consumers.

Many of those luminaries mentioned thus far in this treatise have understood the unstinting contradictions inherent to our species with respect to ourselves and all other creatures. One of the great, but little acknowledged geniuses in this broad, interdisciplinary field of enquiry was the Hohenstaufen Holy Roman Emperor Frederick II (1194-1250), King of Italy, whose coronation in 1220 assured him of a thirty year - thoroughly well financed monarchy - during which time he probably implemented more natural history experiments (some of them conducted on human subjects, including disembowelment to study digestion; raising children – all of whom died by the age of two, in absolute silence to see what languages they would acquire, if not that of Adam and Eve's), proposed more hypotheses and tested more theories and instincts than any other single individual, including Charles Darwin, or even Leonardo. Among his many fascinations were giraffes (which he rode over the Alps) and falcons, about which he wrote one of the first books. He could be ruthless, even insanely so. But he also recognized a transformation taking place in the human species. Some have said he was the "first European" (meaning of a modernist disposition). But underlying his many superlatives and attributions was an obsessive quest to understand the languages and behavior of other species, and where our species fit in that mix. Ultimately, he seems to have wondered what – if anything – our species' destiny boded. He was not alone in that pursuit. [*1]

"Tongue Tied"

Evolutionary linguistics is one of the most hotly contested ongoing debates in all of science. From Aristotle to Rousseau; from Charles Darwin and Max Müller to the Linguistic Society of Paris in the 19th century; from Noam Chomsky, Marc Hauser, W. Tecumseh Fitch, to Johanna Nichols, Simon Baron-Cohen, Daniel Everett, and John Searle, the morphing of two vowels into three; the manipulative complexities of the Neanderthal and *Homo sapiens* tongue, in regards to speech, social or otherwise, the role of such brain areas as Broca's and Wernicke's, the Primary Auditory Cortex and Supramarginal Gyrus; our auditory capacities; relations in thought to our need to communicate…. No one, as yet, knows what combination of morphological, ecological and emotional exigencies made language an act of survival, or of joy. Or how our thinking in words connects us to nature, or, conversely, alienates us.

Proof, therefore, of our communication compulsions are merely conjecture. Said to make us unique (at least according to Aristotle and Claude-Lévi Strauss). Perhaps, in the wake of all these documented historic volitions and initiatives - feats of the imagination that have survived despite so much conflict and stress – the record of our having walked the Earth, felt the Earth, ineluctably invoked consequential challenges that were met - signing to others we trusted – has by force of its own urgings - commencing just minutes and days after birth - engendered an inescapable reinforcing confirmation bias that leaves us with no other paradigm for attempting to see and understand the world, than by communicating with one another. This allocation of portions of our brain is further enchanted by that small percentage of individuals who, to the extent thus far known, experience moments of intense psychokinetic multivariable linguistic ontology at any number of sensory nerve-endings. This semiotic genesis is now part of an initiative to better decode the genetics modulating synaesthesia at the Language and Genetics Department of the Max Planck Institute for Psycholinguistics in Nijmegen, Holland.

We know how culture in human beings has succeeded in passing down traits of interest and predilection, if not outright predisposition; and how phenotypes have evolved in relation to community, geographic isolation, random mutation and the slow accumulation of experience within a singular ancestral clade. Still, none of this looking back can quite elucidate Lascaux, or the Burgos sites throughout northeastern Tamaulipas state, in Mexico.

And no physiological correlate helps us to understand, for example, the famed mid-12th century **St. Albans Psalter** (including 40 illustrations from the life of Christ and four other distinctive sections) with its remarkable illustration of King David in the role of Orpheus playing a three stringed rebeck (viol) for two eccentrically painted ungulates, a goat and a sheep, with a mid-flight pigeon whispering into the King's ear. What was that whisper? What is the secret ingredient carried by the bird? [*2] And does that seminal moment in the history of art and religion suggest that our in-kind communiqués are of less significance than those transductions we have tuned in to between different organisms? Effected participatory interest? Suckled and reared members of other species as we would of our own? Transcended the boundaries within our own neighborhoods, upon our bodies, between the hours of any day or night?

It is our belief that the empirical and residual bodies of evidence all point to the obvious conclusion that other species taught humans how to speak. That they elicited in us an excitement to share, to make welcoming, as well as warning signs; to engage in the expression that the world herself is so clearly devoted to in every biological sense. That ecological situations, one by one, toppled our insularity and gave us to understand the rudimentary requirements of interspecies communities.

Of the twenty so called essential amino acids in our bodies, at least nine of them carry out the building of proteins, and sustain neurotransmission and biosynthesis, fighting depression, combating stress, to various degrees. This has been as physiological fact inherent to some eighty billion *Homo sapiens* who have been born and died. Stress may be key, as it was in the Arctic, where the proliferation of biodiversity under conditions of adversity witnessed a boom several hundred million years ago, lending some credence to the notion, as identified by philosopher John Dewey, that "friction" is essential to creation.

"Apperceptive Nature"

In spite of our primordially colossal multi-generational memory, with all of its diverse cultural and ethical trappings, and in addition to so much discomfiture and misfortune, great dreams, dashed and to various degrees fulfilled, somehow, at no precise juncture likely to be deciphered – a sixty-thousand year old tiara of dried flowers on a buried human skull in the Zagros Mountains, perhaps, a first documented instance of what the ancient Greeks termed "synderesis," a moral consciousness - our expressive impulses began to fix inwardly, cryptically.

Inwardly, as an adjunct, or prime mover of outward expression, our language capacity evolved co-dependent upon the rest of the world. History has proved our evolution from a state of complexity absorbed by unspoken adjectives and impulsive nouns, to the declaration of adverbs in response to both inward feelings and unexpected impacts from outside.

Apperception – our consciousness of our own thought processes - commenced an inward journey compounding the rigorous odysseys of our quotidian fare such that human philosophy entered the world, along with the philosophies of Others – other species, and the individuals among those species who we fully credit with synderetic predilections and realized vision no less (and possibly far more) advanced that that within our own species; seeking to render whole the penumbras and fragments, rainbows and untimely deaths that swept over our hearts and minds. In other words, at a certain point in our becoming who we are today, we found ourselves invested in the communication with other species. The aesthetic corollaries of this conundrum are made abundantly clear in an image on page 121 (as well as the entire back cover) of Sebastião Salgado's recent masterpiece, **Genesis** (Taschen, June 2013). It is the hand of a marine iguana (*Amblyrhynchus cristatus*) in the Galapagos, Ecuador. Yet, it has the shimmering silver mesh perfection of a Spiga necklace by Bulgari, or some serpentine-like electroplating platinum, snaking its perfect way, in this case to a rock, but just as easily to the gold and citrine parure, like that once worn by Joan Crawford.

To find among them, and through them, proverbially, meaning. Derive all shapes, sounds, ideas and inspirations; every Madonna, stained glass window in a great Cathedral, each sunrise and sunset, every mottled shade and soothing creek murmur. Nothing of this magnitude, this naturalism, was ever linear, most importantly, that ephemeral line in the sand. That moment when a hummingbird arrives with precision inches from your gaze and hangs there, a million miles a second, looking forthrightly into your eyes, then disappears in search of that nectar flower at the beginning of time.

Such revelatory instants give rise to naturalism in the from of synecdoche, the mother of human inventions; metonymy – those insights whose body and breath come directly from Nature, filling us with metaphor and every example we might imagine to help us better grasp those questions that remain fixed in our minds, as upon Paul

Gauguin's great canvas, painted in the Tahitian commune of Punaauia, a suburb of sorts of Papeete, in 1897: "Where Do We Come From? What Are We? Where Are We Going?" Today it hangs in the Boston Museum of Fine Arts, but to look at it is to be instantly reconnected to French Polynesia, particularly the Marquesas.

"The Reciprocal Promise of Anthrozoology"

Hence arose the intuitive libraries-worth of anthrozoology amongst our brood. A biological pact, egalitarian, whose unlimited beneficiaries of this and future generations would inhabit the social contract itself, a communion entrusted to the full panoply of living creatures. The very essence of metaphysics, such as we have endeavored to flavor and instill the word.

In his groundbreaking book, **Theory of Meaning**, Jakob von Uexküll wrote that [Animals] "never enter into relationships with neutral objects. Through every relationship the neutral object is transformed into a meaning-carrier, the meaning of which is imprinted upon it by a subject." He was paying tribute, by this deeply-felt manifesto, to Thomas A. Sebeok who had declared that "Zoosemiotics" is a term coined in 1963 to delimit that segment of the field [semiotics] which focuses on messages given off and received by animals," in his work, **'Talking' with animals: Zoosemiotics explained**. [*3]

This explains so much of our own evolutionary penchants; for shade, arboreal, verdant, thickly-clustering liana; for the sound of the hollow sea shell, the alpine Chinook, or African simoom; the Alpine Föhn, the Emperor Penguin's katabatic winters; of sultry languorous nights in summery Maryland or Haiti's Plaine de La Cahuane, the tidal or horizontal waterfalls at one tip of Ilha da Inhaca in the Mozambique Channel, or the subsumed auguries of crickets trilling sensuously around Walden Pond; of pearly micro-cascades throughout Scotland's Trussochs, and the far sound of avalanches in the Kafkas (Caucasus); of the distant eerie joyous thunders on the Pacific, at Sand Dollar Beach in southern Big Sur, accompanied by the acute hints of marine spray that defines shorelines and whitecaps from the Andamans and Nicobar, to the Ruggedy Range across from Whenua Hou (Codfish – the Rakiura archipelago) in New Zealand. Of all those first moments of the beginning of the monsoon, traditionally around June 15th, and the full 180 degrees of perfect rainbows, undiluted by scientific analysis.

Such reciprocal promises –beauty, natural science, art, and the elusive premonitions of the Delphic Oracle, are hallowed in nature, hallowed in the human gravity, and have perennially given us to recognize and cling to that strange and lasting appreciation of other species, principally each and every individual Ambassador for those species, as if the encounter, the received wisdom were perpetually renewed, as during every first day of Spring, as avowed in Chaucer's "Nun's Priest's Tale," sans perturbations, only love. [*4]

During this aesthetic evolution in human beings, dogs as well as zebras (the latter with quite limited success) were domesticated by our kind. Cheetahs, too, from Egypt to Persia. All of the known anthropological assortments of kinship, clan, community structure and self-importance were given loft; one experiment after another in social hierarchy enabled, just as our own inner fixations were breeding ideas that came encoded in more rarified ideals.

In a world populated by so many enigmas; and by vast flocks, herds, pods, gaggles, coveys and broods, but also by the lapidary solitaires – from male tigers to a moose - who clearly exhibited analogous features, desires, fears and capabilities with ourselves, in much more abundance, to be clear, who were we to initiate bold queries of an ontological nature? Paleolithic experience suggests the first hint of a known separation of essences, cognitive dissonance making out a map of shifting – nearly seismic – change: that to which we had belonged without the least doubt, suddenly – at least by any geological yardstick – presently demanded answers to unknown phenomena. Every history of our kind re-examines this parting of ways – like continental drift, such was the drama inherent to a saber tooth tiger leaving go the emerging hegemonies of man – and characterizes this disassociation by any number of evolutionary noumena beyond our direct sensory grasp.

Petroglyphs, entomoglyphs (such as those recorded by Montana mycologist Larry Evans), arborglyphs, the tells above the Dead Sea, the middens of San Nicholas Island… all these reminders of our imprinted equations of dependency upon the Other. The most recent and spectacular discovery (October 28th, 2004) of this profound evolutionary matrix comes from the Indonesian Island of Flores, a cave (Liang Bua) in particular wherein *Homo floresiensis*, standing about 3 feet tall, having arrived on the island approximately 45,000 years ago, was discovered in tact. His bones lay near to an extinct elephant genius, Stegodon, as well as Komodo dragons and huge, now extinct rats. [*5]

In small clans, hamlets, and looking out from vouchsafed citadels across an increasingly changeable world of advancing and receding glaciers, shifting water tables, and the encounter with other hominids, philosophy must have grown up wearing an uncomfortable apperception of fears and hopes. The myriad of great Unknowns we felt about ourselves and our loved ones; populated by so many Others around us, unleashed a quest. Survival may not have been key. Compassion might well have dominated the prevailing instinct of our kind.

Moses, Homer, Hesiod and Thales; Heraclitus, Pythagoras and Parmenides; Lao Tzu, Mahavira and Buddha. The Essenes and Todas. Throughout our book **God's Country: The New Zealand Factor** (2011) Jane and I have elaborated at great length upon these human vegetarian clades. In no certain time-frame an accumulated body of introspection laid claim to systems of access to an elaborate Chain of Being, or teleology, as it came to be thought of; calendars and depictions of the Cosmos that reflected both the familiar and totally unknowable. Deities had their day. This growing body of data, an ecosystems-related approach to introspection, landscape sized systematics invoking human language, bound to gravitate towards that which seeks to protect what is familiar. And, eventually, unfamiliar.

If compassion was, in fact, a central feature of our survival, as many biologists now suspect, along with commensalism, symbiosis, and other mutually-beneficial relations throughout the entire panoply of Nature, it is frightfully clear that the expression of our fundamental need to love Others did not get easier over time, but has been challenged by an increasingly uphill struggle. Why that is will forever remain obscured in the annals of all that is tragic, uncertain and pending, in terms of the human legacy in relation to all the Others who co-habit this planet.

The Jains, particularly as exemplified throughout the life of Mahavira, have contributed a striking and original insight into the interdependency that defines this abiding unrehearsed choreography: humans, along with every other species.

"The Extraordinary Legacy of Mahavira"

Ancient and Medieval texts commending Jain forms of asceticism as a precursor to achieving ultimate liberation from attachments and desires, as well as reaching that unique state of omniscience characteristic of the 24 Jinas or Tirthankaras, is thematically and critically predicated upon the notion of ahimsa, non-violence (or non-interference). It is, for Jain tradition, what we would characterize as "existential biology."

Mahavira, also known as Vardhamana (599-527 BCE), is most commonly cited by Jain scholars and adherents as a man whose more than 40 year odyssey best exemplifies an approach to ahimsa, a very divining rod of all those personal predilections that lend great inspiration to all those who have sought to embrace similar ethical commitments in their own way, in their own time. It is a most difficult road upon which to embark; nor is the culturally enshrined, community-driven Jain ethos of non-violence by any means the end-all. In countries outside India, many young people acculturate at fast-food outlets, consuming meat, a 180-degree reversal of Jain ethical renunciation of all occupations, intentions and actions that harm other sentient beings.

Moreover, the Jain definition of "sentient" pivots upon a most scientifically-oriented psychoanalysis of neurophysics, allowing only for the consumption of creatures with no more than one sense – as perceived within the broad Jain metaphysic involving the dialectics of suṣamā-Duḥṣama-(utmost happiness versus unhappiness); of Syādvāda, or conditioned predication; [*6] and Nayavāda, viewpoints and partial viewpoints; - rejecting harm to the 2, 3, 4, and five sensed organisms, as most vividly portrayed in the 12th century Jain cosmological text, **Sangrahanisutra**, composed in 1136 by Shrichandra, which delineates iconic forms of violence in nature and, by extrapolation, human cruelties.

Because the "fork" may well be the most toxic tool ever wielded by our kind, Jain recognition that the breakfast, lunch and dinner table are the most multitudinous intersections of human violence and the destruction of other species is nothing less than genius; transcendental hospice; the precursor of that long paradise tradition that culminated with all the great dreamers like Hubert and Jan Van Eyck, Dürer and Monteverdi, Vermeer and Beethoven.

It is also self-evident, perhaps, to those who care but are afraid. To those for whom neutrality is safer.

Jain monks, principally Swetambara (white robed) and Digamgara (sky clad), number in several thousand, male Digambara (who wander naked their entire lives being in the extreme minority, in terms of numbers) and secular Jains may number twenty-million worldwide but all would concur, no doubt, that – as both Mahavira, and his younger contemporary, Buddha suggested, absolute non-violence is impossible. Digrambara male monks comes as close as any human beings in history to exerting the least ecological footprint imaginable, as well as in practice. [*7]

Gandhi himself – raised by a Jain – reiterated this reality by stating that while non-violence limps, it is the only way.

In terms of human rights and across all ecological criteria, quality of life and sustainability indicators, India is deeply challenged in spite of Mahavira's legacy.

The question, then, is not what relevancy to today's times can be discerned in the exemplary lives of a Mahavira and countless subsequent personalities and traditions - from Cervantes' ultimate dreamer, Don Quixote - to such original thinkers and vegetarians as poet Percy Shelley. The more troubling query concerns the degree of specific traction in a morally jumbled world; of ethical priorities amid so many competing altruisms. How, for example, does any "statistical" individual in North America deal with the news that over 230 endangered species in the coastal portions of the United States, for example, are at great risk of extinction due to climate change and rising seawaters. How can this looming calamity – one among so many others – be remediated by individual action or willpower? Does our ability to protect and nurture in and of itself offer the least guidance in so complex an instance?

Is there some honed and true pathway that can, from spiritual or artistic injunctions rectify, remedy, offset or prevent today's global, environmental crises? Does the Jain position on non-violence simply echo a hollow refrain, however inspired, or does it provide the true rudiments of some remedial and revolutionary framework for addressing 21st century biodiversity degradation in the context of today's all-out war being waged by the human species against the Earth?

A cursory examination of ancient and Medieval Jain art, sculpture and temple architecture across India triggers internalized resolve, at which point human nature is all that is left to be gambled upon, much like the ambiguities attendant upon the Sabarmati River clean-up project in India's Gujarat state, that has provided a new ecological vision, but only at the expense of thousands of impoverished Gujarati's living on the banks of the river. The gazing eyes of the various Tirthankaras, like the famed "Seated Tirthankara without attendants, Rashtrakuta-Chalukya transition, 10th-11th century A.D., Kambadahalli, Mandya district" [*8] allow for no window on the gaming rules of that nature; no guidance whatsoever, other than inward-dwelling reflection.

It is the same with every work of art, whether Paulus Potter's "Punishment of a Hunter" (1647, Collection of Empress Josephine, Malmaison, The Hermitage), or the penetrating self-aware portraits of Rembrandt. Even the deep movements of joy and empathy enlisted by a Bach's "St. Matthew Passion" are no preamble to certainty. It is well known that Beethoven's Ninth Symphony, particularly its 4th Movement, has been embraced by most of the world's listeners, and that included the Nazis, as brilliantly elaborated in **Beethoven's Ninth: A Political History**, by Esteban Buch (Translated by Richard Miller, University of Chicago Press, 2003).

It was Marcel Proust who somewhere declared that while we can make all the promises of trust and loyalty, of friendship and love in the world, the only thing we always "obey" is "pain". The aggregates of pain across the planet; those specific arenas in which the most amount of pain – read cruelty – is meted out to the most number of neurologically sensitive creatures becomes a sure measure of how we must come to view the influence of spirituality and art, of philosophy and every conceivable metaphysic, as a potential interface with the real world, and to what extent causal volitions may exist, as opposed to expressive Nothingness.

This is the ecological existentialism at the roots of all protection. It presents a special difficulty in terms of human conceptualization and our capacity for formulating appropriate balms – both material and emotional concurrences that are effective pain relievers, such as the ancient Chinese medicinal Corydalis tuber (*Corydalis turtschaninovii forma yanhusuo*) that has been studied by the global Herbalome Project (CHM) and has revealed amongst the some 470 species in the Papaveraceae family to promulgate universal characteristics of what has been described as dehydro-corybulbine, or DHCB. [*9] It rings of density but, in truth, represents everything we care to turn away from – pain. All that horror the Hippocratic Oath was intended to address and cure, not ignore; the massive assault upon life which is our legacy and which we have consistently disowned, whilst procreating full of want, absent reason.

Of course this strikes of some cynicism, in spite of the tens-of-thousands of fine doctors and researchers struggling to enlist increasingly effective modalities of palliative care for all mid-to-large sized vertebrates.

Still, the cynicism resides as a heavy weight on the heart, for it is there, real enough. There be no tomorrow for the hundreds-of-billions of persons we torture and annihilate, the animals who are no different than ourselves. This ultimate zoosemiotic reference to that desperate spate of typhoons and hurricanes of predation, tsunamis of lameness and oblivious death, drought and habitat devastation, fires and heartbreak that kills people, yes, but in far greater number alters the biological history of countless individuals, boom-and-bust generations amongst all other species who die quietly in the dark, unheralded.

Great poets, all sentient beings have some sense of this. Consider the 20 known species of Bowerbirds in Australia and New Guinea who spend so much considerable effort making spectacular nests. In the end – does it really matter? Are they abetting evolution or expending so much energy in spite of it? Was World War 1 part of the evolutionary game plan, or were the dice strictly rigged? Does biology admit to the arbitrary? Was there purpose to Don Quixote's method? Does it render a service, at least from any ecological perspective, to even assess the points and counter-points of any aesthetic? Or is all biological supposition tendentious, irrelevant, in the face of great art, such as Cervantes has so generously, selflessly, given us?

That gift is not in vain. It is the compassion that is a template for all relations between species; the dream that gives us the moral duty, indeed the privilege to converse with Eugène Boudin's cows, frolic with Thomas Sydney Cooper's boisterous goats, stroll musing alongside Monet's haystacks, or wander through a jungle marveling at army ants.

This is not a *system* by which we might assume the conversion to a type of systemology. Throughout the history of philosophy, discussions have weighed heavily upon every conceivable doubt and angst, question of mind, existence, logic and illogic, rights and responsibilities, morality and immorality, perception, false, perception true. One could readily delve into the works of the prolific Bishop George Berkeley (1685 -1753), the famed Irish philosopher who at the age of 25 suggested, in his **Treatise concerning the Principles of Human Knowledge** that perception comprised the full and only possible connection worth speaking about to that environment outside ourselves; that one's mind somehow counter-intuitively held the world in its palm. While it was an eloquent argument against the materialism of his age, it accomplished something unexpected, in terms of men like Don Quixote: epistemological solipsism, as it has easily been denominated and railed against, or extreme subjectivism, which might just be the only defense of art for art sake, in the face of humanity gone awry. Indeed, the personal perception of arcadia is worth more in one fell swoon than all the dialectics of a lifetime.

"Tilting at Windmills"

James Fitzmaurice-Kelly, in January of 1905, addressed the British Academy to commemorate not only the 1605 publication of Part One of Cervantes' unprecedented achievement that was his novel **Don Quixote**, but to point out that (despite slight changes in the 17th century calendars of England and Spain) - both Cervantes and Shakespeare died the same day, the very same hour and the Sun in both countries was rhapsodically said to have "perished out of heaven" April 23rd, 1616. [*10]

By 1616 **Don Quixote**, in full - both parts One and Two – had swept over England. Book One had arrived without initial fanfare in a Brussels edition in 1607. In France, 1614 saw an edition of Book One. The Spanish Part Two, 1616 was published in English in 1620, both parts in Italian in 1622. Such is the migration of fine words, of the loyal tenacious reality of poetry, that it had ignited a continent, though Germany was slow to catch on – no edition emerging before1648 that we know of. The first English translation, by Thomas Shelton would be re-issued again in 1652 and again in 1675. [*11]

But the first 1607 Brussels edition had wielded its cathartic appeal, rapaciously schooling an unexpectedly broad readership across England in large measure due to Chapter 8, Part One, wherein Don Quixote – great dreamer that he is - combats a windmill; bethinks he all those standing windmills across the brazen, smitten plain leading to Puerto Lapice, horrible giants; monsters transformed by a wizard named Friston; a man who was a magician, his full name El Sabio Frestón. Cervantes invests him with a smacking plausibility: he steals books – all of Don Quixote's library; and he has the ability to literally enchant windmills. This is a critical distinction: Don Quixote senses that there is at large a chicanery, this heinous transmogrification. He is not simply delusional –the pratfall preposterousness that many original readers took away from the book - and this makes a huge difference if one cares to analyze the evolution of meaning in Don Quixote. The "Don" is not hallucinating, but putting two and two together. He has mapped a world of morality. Virtue in the hands of this hermit spells one of the earliest case studies in both idealism and humanitarianism.

The "tilting at windmills," which had become a phrase for all times, repeated constantly from Fletcher to Ben Johnson, was actually a well-thought-out response to a very real world of people who did terrible things. Windmills were necessary for life support, as were books. Both had become endangered and Don Quixote, Cervantes, that is, had set his sights on righting a terrible wrong.

"The Evolution of Quixotic Iconography"

The visibility of this storyline was undeniable and by 1620 frontispieces have become part of a de rigueur iconography in which the Knight Errant, wears a "basin on his head and a lance-pennant (as Knight of the Lions") and his Squire, Sancho, bears "a whip and his sword. In the background, a wind-mill" [*12]. By 1648, the Frankfurt edition published by T. Matthiae Götzen had five iconic images, nine in a French edition two years later, and 38 splendid images by 1672/73 in Antwerp. Another edition in Madrid in 1674, was clearly influenced in its pictorial treatment by contemporary trends in Italianate landscapes verging upon the deceased Dosso Dossi, Veronese,

Claude – replicated again in Rome and Paris in editions from 1677 and 1681. By 1687, with the Thomas Hodgkin publication in English, translated largely after Shelton, but with some twists by none other than John Milton's nephew John Phillips, there were for the first time an English-related **Don Quixote**, 14 copper engravings, anonymous, but much like the very original images of 1648. [*13]

All of the 17th century images were stiff and stilted; much like the common interpretation of Cervantes' work at that time: a comic, course satire. A work of foolery meant to dispel myths that might, otherwise, take seriously those silly archetypes of an already bygone era; a comical genre, spoofs not unakin to slipping on banana peels of history.

It appears quite clear that critics from Spain to the Continent appear to have missed the underlying sobriety of Cervantes' achievement, a fact exacerbated by various pirated editions expounding upon the buffoonery under one pseudonym or another. By the late 18th century, Don Quixote might be difficult to disengage from the other much popularized life of Hieronymus Carl Friedrich von Münchhausen [*14], the German storyteller who had taken up arms with the Russian military in the regiment known as the Brunswick-Cuirassier and fought in two wars against Ottoman Turks. His tales got larger by the decade. By 1785 author Rudolf Erich Raspe published the English edition of **Baron Munchausen's Narrative of his marvelous Travels and Campaigns in Russia**, also known as **The Surprising Adventures of Baron Munchausen**. By the late 19th century there would be well over 100 editions of this work, with illustrations by no less an artist than Gustave Doré. Other flamboyant adventure tales were widely disseminated, in the vein of William Combe's **The Three Tours Of Dr. Syntax In Search of the Picturesque...of Consolation...[and] In Search of a Wife, with hand colored aquatint plates by Thomas Rowlandson**; or of the prolific illustrations of Dickens, of **Life in Paris**, etc. by George Cruickshank. In the case of the famed Dr. Syntax, his character was a schoolmaster/clergyman whose misadventures during a holiday tour were among the endearing and delightful tales of woe in the 19th century, a bit like the 20th century "Honeymooners."

But with the engravings by Gustav Doré for a French edition of **Don Quixote** in 1863, an English edition in 1868 [*15], a new tonality of utter magnificence suddenly and irreversibly enshrined the Cervantes legacy. Doré would illustrate the Munchausen adventures in an 1866 edition, but there is no likeness whatsoever either in genre or sensibility to what he accomplished for Cervantes. Doré's 118 illustrations for Don Quixote revealed an extraordinarily erudite and passionately focused archaeologist at work in the trenches of emotional nuance – a pictographic Michel Foucault, George Steiner or Simon Schama – who had finally, after 247 years of illustrations that never came close to touching the heart of the great epic - unearthed the original Don Quixote mythology and breathed the original and devastating depth into the work, transforming stick-figures and flat-affect into a deeply resonating universal soul.

By the 20th century, the English were as profoundly steeped in the particulars of Cervantes' own personal epic as they were in his resplendent masterpiece. Cervantes had become the protagonist, just as history had become itself, in the present day – a warp in the time/space continuum that had never occurred prior to that time. The pain and tragedy of the author's life seemed finally to infuse every articulation, each sentence of Don Quixote's own epic in concert with his neighbor and friend, Sancho Panza, and their horse and their burro, Rocinante and Dapple.

"England, Austria and World War One"

Queen Victoria's relations in Austria, and the fact that Cervantes himself was jettisoned on what, in historic terms can only be described as a nightmarish journey **beginning with the Author's conscription** into the army of Don John of Austria, underscores England's own tragic plunge into World War One. The parallels are severe: Cervantes lost forever the use of his left hand at the Battle of Lepanto, which, by unfortunate twists led to his capture by sadistic Moorish pirates in 1575, five years imprisonment and serfdom in Algiers, failed attempts to escape, subsequent torture and a sublime, purely Don Quixote-like madness, and ultimate redemption. But then, when all had been made right, Cervantes was again singled out by cruel fate: one day more miserable than the next, we can read into his story; literary piracy, shame, more imprisonment on false charges, poverty and early death. The delivery of this novel to the global market was as afflicted with bad-luck as the great 17th century botanist Georg Rumphius' Indonesian botanical, **Het Amboinsche kruidboek**, or **Herbarium Amboinense** (1741), one of the greatest herbaria ever published, but only after a cascade of personal disasters for Rumphius, his family, and the manuscript.

The point of this schematic labyrinth of a biography is that one of the greatest artists of all time, with more warmth and humanity than had ever been visited upon the psyche of readerships, would suffer inhumanity –the fate of Cervantes; that no amount of words and eloquence could stave off the artist's eventual despair.

The same cannot be divined with regard to the life of Mahavira, who's many images, largely in the myriad Medieval Rajasthani and Punjabi Kalpasutras, are, for the most part, undifferentiating in their uniform characterization of the most recent, 24th Jain saint, a mortal whose reception by all adherents, allowed for scarcely any latitude in its depictive stateliness. The ultimate ascetic.

England suffered her own national horrors during the War to End All Wars (as the phrase was adopted by 1943 looking back at "The Great War" of 1914-1918): 880,000 dead troops, approximately, and an overall toll of the dead and wounded approaching thirty-seven million, sixteen million dead (military and civilians) and twenty-million wounded, by far the most heinous, perverse and preposterous show of human futility in the annals of our species, with but one exception: the human induced Sixth Biological Extinction Spasm which continues to escalate, and whose origins may be said to date to the first large-scale mega-fauna extinctions induced by human manipulation of large vertebrates.

The historical, biographical and ecological analogies cascade. By 1950 algal biomass across the oceans had vanished to the tune of over 40%, while 400 parts per million of CO_2 in the atmosphere represents a tipping point now surpassed by what can only be truly described as a runaway Anthropocene (a term replacing the Holocene given humanity's singular role in its blind endeavor to drive the majority of other vertebrates to extinction) dating back at least to the Neolithic revolutions, sedentary human communities, the enclosing of villages in walls, and the large-scale destruction of other species wherever humans congregated across the planet between eight-and-thirty thousand years ago.

By the many twists of human folly and fate, April 23rd 1616 and June 28th 1914 remain inextricably tied to our conscience, like Prometheus pinioned on Mt. Kazbek, tormented by an eagle. Archduke Franz Ferdinand of Austria and his wife Sophie, Duchess of Hohenberg, both assassinated in Sarajevo. The beginning of World War One, pinned on two alleged individuals, Gavrilo Princip and Danilo Ilić.

"The Madness of Metaphors"

By 1610 most of the Low Countries were dreaming of those windmills on the plains of La Mancha. Shakespeare no doubt read Don Quixote four years before he died.

These are historic facts pertaining to a great writer and dreamer; of Don Quixote who was himself enchanted by the path of virtue, by chivalry and by all those Middle Ages tales of Knights in Shining Armor coming to the rescue of the innocent, of those same victims, and all the circumstantial pitfalls and vulnerabilities that Mahavira himself focused upon. That our species has forever fantasized about Utopia, struggled to map the whereabouts of Paradise [*16], and sought to ceaselessly re-invent the Golden Rules. These underlie the greatest of all paradoxes: our goodness, and the manic cruelty, callous indifference to our consumption, and greed that compete at every juncture for an outcome we cannot predict or commandeer as a species. Only individual choices have the power to engender a World War, or the saving of a single life against all odds, 36 million casualties versus the legendary English creations of "Enigma Variations" (Sir Edward William Elgar, 1st Baronet, 2 June 1857 – 23 February 1934) or "The Lark Ascending" (Ralph Vaughan Williams, 12 October 1872 – 26 August 1958). Williams' suffered severe deafness in later life as a result of his exposure to so much gunfire during World War One, where he had enlisted (despite being over 40 years of age) in the Royal Army Medical Corps. [*17]

The contradictions of our power and humane potential shadow the whole concept of protection, baffling any metaphysic with opposites that comprise the full panoply of human nature. That this is the essential theme with which to grapple in an age of enormous strife, replete with a converging and tragic concussion of ecological stressors, there can be little doubt.

Our beloved dreamer like Cervantes, and – and perhaps more focused still - Mahavira - is relevant to today's environmental crises. The dreams and ideals of non-violence have grown into a science, an art, a biological prerequisite for the future of life on earth.

The specific echoes of Asiatic religions and philosophical traditions as so fully embodied in the legacy of Mahavira as to be especially critical during a lasting period when more suffering is being willfully and deliberately meted out to more sentient beings, all by one species, *Homo sapiens*, than at any other time in our documented, or perceived biological history, going back at least 120,000 years if not all the way back to the Cretaceous-Paleogene extinction event. While there have been localized extinctions, mega-droughts, even large scale landscape-sized trophic shifts in species and population compositions, the Cretaceous extinction was global and most likely the result of an asteroid impact, or multiple impacts. That was some 65.5 million years ago. At that time vast percentages of Orders, Families, Genera and Species across the full taxonomic suite of terrestrial life forms died out within a very short time-frame due in large part to the dust clouds cutting off as much as 20% of the sunlight hitting the earth, thus precluding normal rates of photosynthesis.

Later, as if to add insult to injury to Bishop Berkeley's theories, the Paleocene–Eocene Thermal Maximum (PETM), also known as the "Eocene thermal maximum 1" (ETM1) – between 55.8 million years ago and 170 thousand years ago - raised global aquatic temperatures by at least 6 degrees Celsius (10.8 degrees Fahrenheit), extirpating a considerable constellation of benthic communities, including those in foreshore tidal-pools, beyond continental shelves, and into the deepest trenches of the oceans. [*18] The mammalian terrestrial benefits for biodiversity were, oddly, spectacular. And then we arrived on the scene.

The fact that extinction has entered the working vocabulary (e.g., "Transformers [4]: Age of Extinction") underscores the severity of our inured status with regard to acting upon knowledge, or even feeling compelled to mend the world, during a state of unprecedented crisis. Our perceptions are fogged. Our more than five thousand living languages, and all the data pouring in, are apparently insufficient to trigger any rational species-wide response: a mental and/or emotional deficit with inexplicable consequences and an utter absence of technology and consensus mechanisms (or DMM - the CITES conferences Decision Making Mechanisms – for halting the sale of confiscated African elephant ivory tusks, for example) has left us at the mercy of a personal response, both in duration and scope. [*19] But, again to cite Berkeley, those perceptions are our salvation, if we believe in them enough to act according to our heart. This is not a theory.

This is the essence of the metaphysics of protection we are proposing: a means of combating the inevitability of global flux, constant tragedy everyday, somewhere in the world, and of retaining sanity in these circumstances, however one's philosophy chooses to characterize them. The earlier referenced fly that lives as much as a month means that one has four weeks or less to show interest, if not outright love of that fly. Extend the same love with undeviating, undifferentiated warmth to all living beings. It is very easy to do so, regardless of your view about eradicating smallpox or measles, malaria or Dengue fever. 100 million other species will be sufficient, if you set your mind and heart in a dialogue of liberation towards these trillions of Others. The numbers are not daunting. They require no refresher courses in calculus. They are simply, wonderfully - opportunities.

"Despite 250 Million People Murdered Since the Time of the Renaissance"

Our dispositions and pre-dispositions have been greatly debated and debased with respect to the violence enshrouding our species' history, and this deficit currently raging throughout every human community. Some paleoanthropologists have argued that we are, fundamentally, a dangerous mammal, all other species coming near us only at their peril. Others have been far more generous in their characterization of *Homo sapiens*, going so far as to adduce a most beneficent and universal craving in our hearts for peace and love and the ability of our kind to extend olive branches of compassion and tolerance to all concerned: either in practice, or certainly as an ideal. Between the many brutal truths that belie our capacity for nuclear weapons and cruelty, and a goodness that betrays a far less draconian indictment of humankind, indeed, an actual history of good deeds, Ashoka's Edicts, Ten Commandments, inspired international legislation, and the evolution of moderation and care, we see everywhere grievous contradictions.

Note the extent to which the majority of the world was involved in World War One: an Armenian genocide in 1915; Ottomans attacked the Suez Canal; British forces were struggling to capture Baghdad; Serbians overtook Macedonia; the King of Greece had resigned while troops in Slovenia, Croatia and Bosnia fought it out. On the first day of the Battle of the Somme, 19,240 British soldiers were slaughtered, and another half-million Englishmen would die during that Somme offensive. The largest war of battleships in history took place in the North Sea off Jutland while multitudes of Turks were being compelled to evacuate the southern Caucasus. As Gandhi failed to make his voice heard to the British, nearly 48,000 Indian soldiers forced into this conflict were killed. Tsar Nicholas II would soon abdicate his throne as Russia plunged into a kind of unholy madness of empty revolutionary refrains; volleys of tragedy that echoed from the horrendous Battle of Verdun to the delays in peace negotiations by the Entente forces which refused to talk peace until all those parts of Europe that had been occupied - a word of some resonance today - were evacuated by the Central powers. By the Spring of 1917, President Woodrow Wilson had pushed Congress to enact the Selective Service Act which quickly drafted 2.8 million young Americans into the conflict. And so the narrative, that concluded with an Armistice Day, November 11, 1918, -so well known, yet so easily forgotten by later generations- continues, from northern Iraq and Syria to Nigeria and Gaza. [*20]

Yet, in spite of the horrors of a World War One, and the lessons we have supposedly taken from those terrible trenches of death, some elements of the human drama have not only remained static, but many have gotten worse.

"Perpetual Peace"

In Russian Kaliningrad, once part of Germany (Königsberg), a plaque on Königsberg Castle is inscribed to her most famous native son, Immanuel Kant (1724-1804). It reads, in part, "Two things fill the mind with ever new and increasing admiration and awe, the more often and steadily we reflect upon them. The starry heavens above me and the moral law within me." During the final years of this great philosopher's life he was deeply and persuasively absorbed by the relationships of morality to practical matters of everyday human life, in other words with that aspect of the human narrative that has never changed: Utopia versus the all too narrow, stinting, merciless if not ruthless mundane tidbits of life; good versus evil; idealism versus pragmatism.

Among Kant's books, **Perpetual Peace** (1795) and **Metaphysics of Morals** (1797) stand out as utterly original masterpieces, holding forth a kind of Talmudic, steadfast belief in, and insistence upon moral imperatives. Perpetual Peace was one of the most clearly reasoned appeals to the heart, to idealism, and to systematic logic ever conceived, certainly since the time of the great Rabbi, Hillel the Elder (110 BCE, died 10 CE).

Others, like Ayn Rand have questioned this reasoning. Ian Johnston has formulated an overview of the debate. In writing about Kant with reference to Plato's character of Thrasymachus, from the opening pages of the **Republic**, a very different ideology had arisen, a myth of self-perpetuating interest, namely, that the Greek Kratos, or Political Force, clashes with Ethos, or Moral Behavior, and that justice can only be established by those with strength; that "might makes right; that we have no "moral obligations," as Johnston comments, "only imperatives of power." [*21] Johnston quotes a passage from Perpetual Peace that beautifully summarizes the philosopher's prevailing concept, namely,

"Even a continent that feels itself to be superior to another, regardless of whether or not the latter stands in the way of the former, will not fail to exercise the means of increasing its power, plundering and conquering. Thus, all theoretical plans for civil, international, and cosmopolitan rights dissolve into empty, impractical ideals; by contrast, a practice that is based on empirical principles of human nature and that does not regard it demeaning to formulate its maxims in accord with the way of the world [i.e., in accordance with universal moral laws] can alone hope to find a secure foundation for its structure of political prudence." (From **Perpetual Peace**)

Hence, the appropriate focus upon two parallel narratives, or discourses, that have served to both alienate and frustrate humanity since its earliest documented records, namely, the search for peace and love that is within the grasp of every child's heart and soul, and the violence all around us. Heaven and hell, and those endless examples, personal experiences, and expectations with which one would prefer to dwell, and towards which aspire. Hence, writes Kant in the Appendix A, "Deduction Of The Categories" in his **Prolegomena to Any Future Metaphysic**, "All cognition requires a concept, however incomplete or obscure; and this, in its very form, is something universal, and which serves as a rule." [*23] By seeing Cervantes as the bridge between Bishop Berkeley and Kant; Don Quixote as the modern Mahavira, we can see our way clear to enacting the liberation of sentience without the least restrictions, qualms unresolved, or impediments.

Again: this is not a system of logic. If anything, it is a fashion of pragmatic reverie; no bio-blitz, but a non-invasive cuddling conducive to science, but – more importantly – to the health and well-being of those around us. Of course, the world being the world, there will be fumbles.

"Ethical Labyrinths"

The ethical cores of human nature are troubling precisely because they represent a plurality of motives. Some things trouble people more than others. What creates a true revolution in one community, becoming a legal standard, a basic right, duty, responsibility or cultural norm, may be utterly missed, ignored, stubbornly denied or even erased from history elsewhere. Half of all Americans are incredulous that there is such a thing as the "Tea Party" of fanatics and historical illiterates who spout more inanities and untruths than have ever been dared or even imagined in American history. For months the French government called for "humanitarian corridors" in Syria before demanding outright that the leadership step down. For years nothing happened, though chemical weapons – after being deployed on Syrian citizens – were destroyed following global support of United Nations demands. Global military relations are one window upon vacillations predicated by fickle power plays. Syria's President has – as of July 2014 – again been "re-elected".

Animal protections are a far more outrageous sham.

For example, in November 2011 a ruling by the U.S. 9th Circuit Court of Appeals "struck down the U.S. Fish and Wildlife Service's 2007 decision to remove [grizzly] bears from the endangered species list." [*24] At the same time, defying the humane decisions of over 85 other nations, the U.S. "continues to defend" the use of steel-jaw leg-hold traps, a hideous throwback to Medieval nightmarish behavior. [*21] The numbers of animals killed under such unimaginably cruel conditions in the U.S. for the year 2011 was 6.8 million. But, as the Animal Welfare Institute Quarterly analysis indicated, "these figures are most certainly gross underestimates, as they were compiled by the Association of Fish and Wildlife Agencies (AFWA) and rely on data from state game agencies, a number of which failed to submit any data at all." [*25]

In the case of standards pertaining to agriculture and bioethics, you have industry-wide pundits suggesting that by creating something that has a low probably of occurring naturally, it is, nonetheless, natural. Others would argue that by breeding pre-existing natural resistances into certain seed types, like wild pepper, that this is counter to nature. [*26] In an horrific action, scores of oak trees uprooted from Thousand Oaks , California were done so because the law there stipulates that owners of those trees who planted them can kill them at will, whether the species is native or not. The complex 2013 U.S. Farm Bill had amendment insertions in the House of Representatives that would undermine the right of states to engender their own animal welfare standards for imported products, either from other states or nations, which further exacerbates the underlying reality that the long-standing nature/nurture dialectic has not yet even contemplated the additional component of pain, suffering and outright cruelty. Nor does the Farm Bill recognize that November 2013 was the warmest on record since 1880, and what the anthropogenic science entails. [*27] An extremely hot summer day in the Arabian Peninsula, say 49 degrees Celsius or 121 degrees Fahrenheit, is likely to become the norm as climate change escalates. Moreover, in many nations experiencing such heat there are still no labor laws protecting workers dependent for food on whatever jobs they can get from having to do heavy construction work in such prevailing climatic conditions.

Half of the world's women and the majority of children are still clambering for their basic human rights, if they even have the energy at the end of the day. A staggering 200 million environmental refugees are projected during the coming decades, as a complex array of ecological degradations, or stressors, combine to compound existing misery and make it worse.

If climate change continues to escalate unmitigated, that figure of 200 million could easily swarm, infecting half the human world, or between 4 and 6 billion people with a fever to escape their local circumstances, but with nowhere to run.

Conversely, the majority of these refugees will live in cities, and cities are fast becoming the place to be: the way to unburden remaining wild or semi-wild regions of the planet. All of the infrastructure, re-engineering, energy efficiencies and molecular genetics will apply by quantum leaps to human urban life. Little surprise that the abandoned Ukrainian towns of Chernobyl and Pripyat have become postradioactive wildlife sanctuaries, liana all but obfuscating the lost hamlets of Dachas and schoolrooms.

"Post-Holocaust Biologies"

We may surmise that some 50-to-100 billion terrestrial vertebrates - mammals with genetic profiles quite similar to our own - are slaughtered under Holocaust like conditions for unnecessary human consumption every year, not to mention as many as 2.5-to-3 trillion fresh water and marine vertebrates who are also murdered out of human desire. These numbers are celebrated with a compulsion as food analysts remark upon the fact that at least 16% of human protein comes from the seas, as if to suggest that climate disruption with its oceanic acidification will likely disturb this otherwise essential system of serving up thousands of species upon the dinner palate of one species. In Dr. Marc Bekoff's recent seminal book, **Rewilding Our Hearts: Building Pathways of Compassion and Coexistence**, he writes "…it has been estimated that, each year, 37 to 120 billion farmed fish, 970 to 2,700 billion wild fish, and 63 billion farmed mammals and birds are killed for food." (New World Library, 2014) The multiplication of so many individual killings from cottage industry to multinational corporation, from community to nation, from household to region, is obviously abstruse. There is simply no way to thoroughly quantify the actual extent of carnage meted out by humans, as hard as such data crunchers at the United Nations Food and Agriculture Organization, based in Rome, might try. Just in the tiny Danish territory of the Faroe Islands, the "grinds" or massive marine mammal hunts have only increased in relation to the number of nations and NGOs trying to shut them down. A human population of some 30,000 managed to massacre an estimated 1,200 whales and dolphins, producing "hundreds of metric tons of meat and blubber." But these are just estimates. And then all of the road kill; the unheralded deaths by recreational fishermen and hunters, not to mention the rampage of poachers throughout the world. The loopholes in CITES and Pelly Amendment infractions. The deliberate "Animal Control" killing of millions of animals each year, just in the United States. The swatting of flies, killing of wasps and snakes and spiders; or flushing down toilets of live small mammals in households throughout the world. [*28] Political, moral, cultural, economic, psychological, emotional necessity?

"Adding Food to the Equation of Ethics"

At any one time there are thought to be about 25 billion chickens in the world distributed genetically across approximately 55 known breeds, all descending from the Red Jungle Fowl from today's Vietnam. The vast majority of them end up being slaughtered in monstrous ways. Thrown into chippers live, tossed about like small sacks of sand, strung up, beaks burnt off, throats sliced open. Chicken intelligence has been much studied of late and their orientation to the world commands esteem, or should, for its neurological sophistication. But the majority of humans think of a chicken only as a slab of meat between two pieces of bread. While nearly 80,000 edible plants are known worldwide, only about 150 species have ever been cultivated in large scale while 90% of most human food derives from no more than 20 species. So why chickens? The answer does not reside in philosophy, ethics, or science peer reviews. It cannot be attributed to some subtle principle of evolution. Is it simply attributable to human custom?

There is nothing about the six gustatory dorsal taste receptors amongst a landscape of micromogul papillae up and down the tongue that condemn us to eat chicken, or beef, or anything else. (The unami receptor remains little understood, though is associated with a "salty taste"). But one thing is discernible about the slaughter: In an average year during the first decade of the 21st century, American companies profited to the tune of $134 billion from the killing of animals or the growing of food to feed animals that would be slaughtered, according to the U.S. Department of Agriculture. [*29] And that's just within the U.S.

In our book **God's Country: The New Zealand Factor**, we examined the consequences to biodiversity of the adoption of a vegan lifestyle. We peered through a multitude of studies for purposes of drawing even the most rudimentary conclusions. Much of the data pertains to the rough number of non-target animals killed by heavy equipment, from amphibians to a wide range of rodents and avians. We were not able to compute for terrestrial invertebrates, let alone any smaller living creatures. The number of attrition animals was 15 per hectare for intensively grown systems; half that for systems involving ruminants foraging on pasture. Researcher Steven L. Davis asked the question, "If half of the total harvested land in the U.S. was used to produce plant products for human consumption and half was used for pasture-forage production, how many animals would die annually so that humans may eat? His answer, 900 million attrition animals in a vegan model, 450 million in a forage production model. [*30] The suggestion that just in the U.S. a vegan model would kill 900 million animals leaves much to be imagined. Some have challenged these numbers, suggesting an agricultural least-harm system that could generate, for example, the "20 kilograms of protein per year recommended for adults" with as few as one wild animal per 3.3 adults being victimized. [*31] The amount of vegetables grown for use in feed to animals who will be slaughtered; the amount of wastage throughout all food-related industries; the amount of water, soil, pesticides, nitrogenous fertilizers that go towards growing vegetables versus animals for slaughter… All of these components are key to better assessing the mismatch between any conceivable ethical system and the human reality across the planet, including the approximately 1.3 billion bovines at any one time, caught out in the nightmare of Concentrated Animal Feeding Operations. Over 270 million tons of meat documented in 2007 in terms of human global consumption; a figure estimated to double by 2050, but given the fact that demographics are far outpacing all statistical predictions, and have continued to do so every year for decades, it is likely that half-billion tons of meat will be upon human plates within a mere decade.

Such facts invite debate, but towards what end, given such profuse numbers that do not even begin to account for so-called non-target clusters of wildlife? Those survivors of this machinery of human nature are either in the last remaining wilderness, which dwindles more rapidly by the minute, or are by good luck vouchsafed along inconvenient fence lines and degraded margins affording the highest probabilities for the survivability of rare and endangered species, as these are the very areas where mechanical devices cease to operate. The debates also touch upon the reality of what could be termed "a one percent solution," namely those same seed sources on the margin lines in countries like Haiti or El Salvador where studies suggest the potential for regenerating second, third, fourth growth forests that can restore some semblance of sustainable nutrient turnover in soils, engender watersheds, and nurture the return of native biodiversity. [*32]

"The Origins of Massive Suffering"

Are there appropriate methods for harvesting earth's generous bounties? Is all that aforementioned destruction simply attributable to market forces gone awry? To trade agreements and their lavish lobbying arms that harm, rather than upholding the integrity of the environment? To cash crop expediencies that undermine other, more enduring options? Farming internalizes systemic fallacies that militate against any consistent human ethic in agriculture. One might date these profit-driven undercurrents back to the demise of matriarchy thousands of years before Mahavira; to the rise of grain surpluses and slavery which occurred co-dependently; even to the invention of writing itself, which we now know to have been one of the earliest, most efficient methods for keeping track of slaves and sacks of grain.

As with all things etymological, phrases like factory farm take on particular poignancy when considered in original linguistic context. By 1560, for example, the French word "factorie" referred to an "estate manager's office," or, in Latin, "factorium," meaning "office for agents." By 1618 the word was synonymous with a mill, a maker, of a "building" where goods were made, as distinguished from something made at home. The word "farm" presages the mechanization process, as it stems also from the Latin word "firma," fixed payment, connoting a big landowner who needed fewer and fewer people to be involved because of increasing levels of mechanization. Farming itself has never inculcated the notion gleaned from Renaissance or Barbizon paintings where only a few gorgeously figured individuals were doing the heavy lifting, veiled in a kind of Romantic or Fontainebleau glory.

"The Invertebrate Quantum"

Don Quixote sought to rescue fair maidens. Mahavira focused upon every living soul. His legacy even seeks to attribute sentience to dew drops. Beyond human agriculture, if we add the remaining 97% of all known life forms, namely, the invertebrates, those without a backbone, or vertebrae - all their habitats, and the as yet mostly unidentified quanta of organisms, we are in a biological and ethical quandary.

Consider just two of the most notably manhandled of all invertebrates, namely, the nematode *Caenorhabditis elegans* (from the Greek, "rod," and the Latin, "elegant"). These free-living microscopic, clearly elegant roundworms are surely not the Vermes as in vermin, that used to be their sobriquet, but, rather, gorgeous creatures of the world.

Then there is the ubiquitous fruit fly *Drosophila melanogaster*, used in more experiments than any other creature known to human kind – known in the alien sense. Others may be added to the list of candidates for easy torture: sea anemones, sponges, and all of the arthropods, mollusks, and annelids. In the case of the *Caenorhabditis elegans*, it is, indeed, a sensuous and exquisite looking being, with its curvaceous transparency and a full body of physiological systems - the same ones found in rodents and people, by and large. Because *C. elegans* lives only 14-21 days, is around 1 millimeter in size, it has become a model organism, as biologists think of it, fed while alive in a petri dish to E. coli, another supermodel candidate for manipulation, by the hundreds of millions, so that scientists can continually study its unique operons, or gene groups. It was the first organism ever to have the full genome sequenced by people - about 20,000 protein-encoding genes and over 100 million DNA base pairs. [*33]

In other words, a complex animal that produces both sperm and eggs, that has a life. How many are there in the world's soil? Nobody knows. Does it dream at night? Is there a legitimate ethical question to be posed regarding our unexamined manipulation and killing of such organisms?

Those who might think it ludicrous to bother worrying about the killing of insects and spiders are, in fact, losing some ground. Mahavira's legacy, and that of every Jain monk, commends all organisms. Arbitrary generosity? Nothing to lose? Or some stellar conversion of a primordial humanity into a new nature?

The 1968 edition of **Invertebrate Zoology** was quick to point out how arbitrary human bias is with respect to its own relatives [*34] notwithstanding that all animals, including flies and spiders, share some 1500 ancestral genes that are animal-unique. [*35] Transmuting the formidable myth of a sapient human species into a non-violent organism, that is to say, a creature that exercises a minimum of violence to others, takes active steps to protect others in order to stay alive, would necessitate nothing so lofty as a total sea change, as many might at first assume.

Indeed, a pantheistic ecology is not merely quixotic. The goal of a universal non-violence does not make beasts of windmills or windmills of beasts. While some might think otherwise - citing the transportation and energy sectors, for starters - the kin altruism, emotive responses, celebration of nature and beauty, reverence for life other than that of our immediate family members (albeit a statistically checkered set of data) - all come down to pre-existing qualities that exist in the large measure of our lives already. We have amply demonstrated a capacity to play with Frisbees, pass out sweets, hug one another, and perform lovely music, not just kill. Our spiritual solace comes from so many sources - aesthetic, spiritual, the quest for harmony – even whilst others of our kind have committed genocide and continue to inflict so much harm and pain. Our propensities may be schizophrenic, but our capacities to volunteer for the good of all, to come together in so many myriads of communitarian association – from the classroom to the town hall to the voting booth - suggest that we are seekers after something more than manipulation, greed, crime and punishment; that we desire to make the world a true paradise, or as close as we can get to it. That urging is an ideal; an impetus after something that, while elusive, is not impossible. Don Quixote and Mahavira were not impossible dreamers, not by any means. Of course, neither did they know of automobiles whose speed and mass on average result in the obliteration of unknown quantities of invertebrates. As for vertebrates, at least a quarter-billion are killed just on U.S. highways each year, which does not begin to quantify the global attrition rate.

"Reinventing the Panjorapor Frontier"

The Jain enshrinment of this quality of the heart, quality of mind, behavioral component of all human life is the sharana, or refuge; the panjorapor, an animal refuge; and the Samavasarana, the "Refuge to All" in which Mahavira, any Tirthankara speaks to all the animals and plants. This motif is also central, for example, to the scene of Saint Francis of Assisi delivering his "Sermon to the Birds" as first transcribed by hagiographer Thomas of Celano in the mid-13th century. This ethological revolution, as we today think of interspecies communications, is second nature to every one of us who grew up with animal and plant friends; climbed trees, nurtured injured birds back to well-being, cared for the family dog, derived great pleasure from kittens, and the like. For children in common around the world, this love of nature is largely undeviating.

By high school, particularly for urban kids in tough environments, nature has become a luxury. The dialogue to understand and halt violence between suburban and urban victims has become one of many tasks embraced by New Town, Connecticut survivors of a school massacre of 26 individuals in 2012. [*36] One poll taken among American teenagers showed that upon graduation the average amount of time these young people had spent in anything approaching a true wilderness was less than 15 minutes; and that while the average teenager could rattle off hundreds of brand names, rock bands, the latest app's for their Android or iPhone, they knew fewer than a dozen species by common name.

On farms throughout the world, the commoditization of animals - many of which the children of those farmers grew up affectionately perceiving as members of the family - those same children, soon to become accountant-farmers with computerized implants to identify each bovine in their midst, have all but lost, in so many cases, the connection to animals, other than as Cattle Beasts, as they are known in a country like New Zealand. Ironically, one of the most popular children's books in New Zealand involves a real life cow, caught up in a real life drama in which that cow, with all but a number, actually saved the life of the farmer's wife during a horrific flood. Such anecdotal paradoxes are legion, as we all know. And as birders around the world doing their own bird counts, Dr. Marc Bekoff has reminded us, such anecdotal material is actually critical data that builds into indisputable science.

But what can science say to reconcile children growing up with animals, loving them as pets, and then being easily persuaded by their parents to collaborate in the ruthless slaughter of those same animals?

The panjorapor is an ancient impulse measured today by its effectiveness in the world of biodiversity conservation, no-kill shelters and unremitting initiatives to diminish the rate at which habitat is being fragmented and destroyed.

"Aesthetic Ecologies"

The Bohemian-Austrian poet Rainer Maria Rilke reminds us that questions are critical. And one question that weighs heavily - involving corporate personhood, and the power irrationally ceded to those with the most money- is this: can a living being be reduced to a "product" in a person's mind? The question has an answer and it is, heartbreakingly in the affirmative. Whether in the form of a lamb chop or a patented mouse, or the thousands of gene copyrights, this living planet is engaged in warfare with one species, us. And this is nothing new. Twenty years ago, when the human population of the world was adding one Los Angeles every three weeks (versus one Philadelphia every three weeks, presently), we undertook a comprehensive global study which was published under the title, **World War III: Population & The Biosphere At The End Of The Millennium** (1994). Today, with climate change accelerating much faster than anticipated, habitat fragmentation, species extinctions and the youth bulge demographically taking more than half the world's nations by storm, the global situation is more dire than we could have even projected back in the early 1990s. It pains us to say so.

Every intellectual discipline has history to examine, and in every field - from paleontology to microbiology; from geopolitics to comparative religions - we recognize the all too familiar precursors of present strife; thousands of years of vicissitudes, trial and tribulation, theory, praxis, interpolation and perplexity, from one graveyard to another, as intimated in all the art forms known to our human presence.

Our paradoxically rich addiction to art forms constitutes an ecological and moral conundrum. As Emile Zola, Jean-Paul Sartre, Sartre's cousin Albert Schweitzer and Pablo Picasso each said in various ways, art that it is not socially conscious, activist-driven, is a "useless passion."

Consider the myriad of expressions that grace the world's museums, like the petite but astounding Brunei Gallery at the University of London, with its Foyle Special Collections containing numerous great Asiatic treasures: the remarkable illustrated Mughal copy of the **Anvar-I Suhayli**, or **Lights of Canopus**, by Al-Khashifi, dated 1593 CE, of the Safavid Dynasty, in turn, deriving from earlier Arabic, Syriac, Pahlavi and Sanskrit tales. Much like the work of Aesop, these are animal fables that speak to a universal outflow of sentiment and ethical feeling that orient towards solutions-based aesthetics. In the case of the **Lights of Canopus**, the 27 miniatures depicted present to modern-day viewers something of a Garden of Eden, an island paradise inhabited primarily by primates, as American art historian John Seyller wrote back in 1986. Seyller focused upon the fact that the iconographic details drifted over time such that familiarity with the return of a said monkey king to his kingdom, and the ridding of the island of its bear population, as well as the famed self-sacrifice of the hero, Maymun, the so-called "patriotic Monkey Luring the Bears to their Fate" in some "fiery desert" were details long subsumed by artistic priorities that should give joy to those who beheld the elaborately illuminated pages of the manuscript. [*37] Indeed, these miniatures foretold of ecological collapse, species extinctions, the crisis of genetic cul de sacs on island geographies, and the ambiguous relations between species.

In 2008, Uc Maymun, translated into Turkish as "Three Monkeys," was Turkey's official submission to the Academy Awards for Best Foreign Language Film. Like all animal fables - from the Buddhist Jataka tales including such a wide-ranging cornucopia of sentient nomads, the world's literature and art, drama, opera, and religions are steeped in various guises of symbiosis and mutually-beneficial means to an end. That end is a unison that, in Jain tradition is best summarized by Mahavira's declaration, as recorded in the **Acaranga Sutra**, of interdependency amongst all living beings, or so it must be concluded, given that the actual text, as interpreted by the Svetambara sect, was not recompiled or edited for some 980 years after the death of Mahavira; nor is it even acknowledged by the Digambara proponents who, instead think of it as the Mulachara, "a Jain text by the 2nd century CE Acharya Vattakera," as John Cort discussed some years ago in his important essay on "Bhakti in the Early Jain Tradition." [*38]

Not unlike the multiplicity of nouns and adjectives within Hindu tradition which elaborate the identity of Krishna as Jagannatha, or Lord of the Universe, in such frameworks as Govinda or Gopala - finder and benefactor of cows - so to Mahavira's influence throughout Jain traditions in terms of his divided interest between asceticism and devotion towards all living beings invites consideration of a unique question stemming both from art, and from theology; a global thematic challenge inherent to all religions, namely, the notion - as beautifully grappled with in St. Augustine's **Confessions**, of humanity's flawed nature, its dualism. Not a "bad nature" but a species of "restless hearts" as indicated in Book 1 of the **Confessions** which references "Beauty ever ancient ever new." Augustine takes this logically to focus upon our ability to perceive God as one who is intimate in his, her, its connection to all individuals on earth. Critics argue that for all of the greatness of Augustinian tradition it is, alas, God-centered. Yet, from a Jain perspective where, in truth, there is no God, but a world of living sentient beings, eternal, without beginning or end (connoting no beginning or end to cruelty or, for that matter, beneficence) and the nobility of non-violence toward them, as acted out on the world stage of ahimsa (non-violence) versus hinsa (violence), Mahavira exemplifies what should be considered a reformer, an activist, a man who knew how to enact peace and convey the legacy of reconciliation of opposites, the end result being the famed Parasparograho Jivanam: three Sanskrit words - paraspara (mutual), upagraha (assistance) and jiva (living beings, jivanam being the plural, of course, of jiva).

This sutra is, indeed, the most notable and self-evident commitment of Jain tradition, and the one most applicable to a 21st century planet, but one that was not formally introduced to Western tradition prior to around 1884, when Hermann Jacobi translated the Jain texts as the "Jaina Sutras," parts, of the **Sacred Books of the East Series** (Volumes 22 and 45), encompassing the **Akaranga**, **Kalpa** and **Uttaradhyana sutras**, as well as the **Sutrakritanga**.

This universal multiplicity was anticipated by the world's first chronicler of different organisms, Aristotle, who acknowledged "isolated species which fall under no recognized 'greatest genius'" [*39]

The cosmic whorl of sayings, attributions, theories and systems attributed to Buddha also feel their way towards a similar realm of moral injunctions within a specific population of vulnerable sentience. For example, in the Pali to English translations of the **Anguttara Nikaya**, or **The Book of the Gradual Sayings of Buddha**, the fourth of the five nikayas or collections in the **Sutta Pitaka**, the second division of the **Tipitaka**, comprising over ten-thousand suttas or discourses allegedly delivered by Buddha and his disciples during Buddha's 45-year career as a teacher [*40], suggests that as in the case of a lion who would reincarnate into Mahavira, the praise conveyed by Buddha in the **Anguttara Nikaya** also went to the lion. The Abhidharma is often thought of as the core study of Buddhist psychological analysis for ontology, the framework for salvation or liberation; the very nation of existence and the mechanism of the mind. It is highly valued throughout Mahayana Buddhist traditions from India to Tibet, from China and Korea to Japan. It was Thomas William Rhys Davids (1843 - 1922) who made much of this Pali literary heritage accessible to Western readers in translation, along with his wife, also a Pali scholar, Caroline Rhys Davids (1857-1942). Among the countless re-affirmations by Buddha of ahimsa comes from the **Anguttara Nikaya** 5-177: "O Monks! The following five trades should not be followed by a layman: - 1) trade in arms; 2) trade in living creatures; 3) trade in meat; 4) trade in wine; 5) trade in poison. [*41]

The convergence of an Aristotelian/Jain/Buddhist ethos is supported on many levels of interpretation that defy precise itineraries of confluence, but summarily suggest a human predilection throughout philosophy and ethics towards this conceptual base upon which to view the biological universe. We know that Asoka embraced Kashmiri Jainism when he was the Viceroy of Ujjain in approximately 260 B.C. His grandfather Chandragupta had also vowed to follow its precepts throughout the Magadha Empire of which he was ruler. In many of the rock inscriptions attributed to Asoka's influence is the revered Jain reference, Deva nam piya, or, the Beloved of God. [*42]

From an entirely different perspective, art critic Christopher Knight, writing in an essay on the "Leonardo da Vinci: Painter at the Court of Milan" exhibition in London during the Autumn of 2011 at the National Gallery concluded his discussion by reminding us that "material substance merges with immaterial spirit, forming a boundless reservoir of charity and faith." [*43] This too tends towards a mellow substantiation of the intimate, personal connectivity that Bishop Berkeley was, in our estimation, referring to. An approach towards dispensing with nonsense – the debate of the number of fairies on the head of a pin – and actively liberating one's own inner reserves, one's conscience, one's unconditional affiliation with all Others.

"The Nests of Yesteryear"

Salvador De Madariaga wrote of the end of **Don Quixote** as that moment when the heroic personage lies dying in his bed, tells the kneeling squire, Sancho Panza to "Leave those fooleries" and "having gained his reason and lost his reasons for living... (declares) 'There are no birds this year in the nests of yesteryear.'" [*44]

Nothing like a bird has so captured the human propensity for discerning our inextricable links to a global hearth. A study in Lubbock Texas of home sales between

2008 and 2009 suggested rather conclusively that a "blue jay in your backyard could add $32,000 to your asking price." [*45]

But such urban birds have an additional complication: they appear to be deemed less important by children, polls of whom suggest them to be more concerned with exotic international species, than those in their backyards. Yet, in combination, urban and semi-urban backyards are part of that global movement in which - between 1992 and 2005, the urban landscape more than doubled to "0.128 percent of Earth's surface," a number that does not account for the more than 50% of human-arrogated global NPP (net primary production, or chlorophyll-producing agricultural and agroforestry areas). [*46]

Don Quixote's "nests of yesteryear" have gone further back in time than the Renaissance in terms of attrition and our nostalgia for the rapidity of change, and the loss of experience. Of the nearly 10,000 known bird species and 22,000 avian sub-species on the planet, tens-of-billions of whom migrate with the excitement and exigency of Renaissance master Michael Praetorius' (ca. 1571-1621) "Terpsichore Dances Part II," it only takes one little bird, a half-ounce robin red-breast (or, more exactly, orange-breast, *Erithacus rubecula*) in a British winter, for example, to transform the life of a child. As John Humphreys has described, [*47] despite her short lifespan (12-to-13 months) she made Keats's "Ode To Autumn," Robin Hood, Beatrix Potter, Christmas Cards and a wintry landscape all come alive, for she remains while others, such as the cuckoo and the martin, have long flown south to warmer climes. Each generation teaches the new one to be trusting. It is well known that robins will come to feed from a human's hand. For unlike her Greek linguistic roots, or erithakos, which means solitary, she is not. Ultimately, this iconic little creature, a member of the flycatcher family in Europe (not the thrush family, as in America) and England's unofficial national bird since the 1960s, she enabled William Blake to utter one of the great clarion calls of all, when -in his "Auguries of Innocence" he reminded us that to put a robin in a cage was to put "all heaven in a rage."

The "nest" is the epicenter of any future metaphysic that would seek to find protective legislation, emotional nurturance, philosophical conservation, activist strategies to save the birds.

"An Ecological Rorschach Test"

Initially, in 1921 when the Swiss psychiatrists Dr. Rorschach and Hans Huber first developed it, and it was then published under the name of **Psychodiagnostik**, the montage conundrum-slash-test it went on to become the most talked about, administered, and scientifically scrutinized of all psychological techniques. It has numerous critics but at least 80% of all clinical psychologists still utilize it. The original test encompassed ten inkblots that combined such images - or the perception of images -

as bats or butterflies, humans, quadrupeds, animal remains, such as skins, and a combination of vertebrates and invertebrates, from the head of a rabbit to a blue spider. Thousands of publications have attempted to ferret out reason and unreason, logic and illogic from the perception of these inkblots.

Transpose the **Psychodiagnostik** into the ecological circumstances of nearly a century later, and through what blizzard of inflictions might we find our true path? A random sampling of ten 21st century "inkblots," so to speak: 1) "Animal Migrations May Be Moving Towards Extinction" [*48]; 2) "Are songbirds arriving earlier in Thoreau's Concord as the climate warms?" [*49]; 3) "Protection is delayed for greater sage grouse" [*50]; 4) "Alexander Gillespie: More effort needed in global strategy to protect birdlife" [*51]; 5) "Value of Small Patches in the Conservation of Plant-Species Diversity in Highly Fragmented Rainforest" [*52]; 6) Pied Wagtails (*Motacilla alba*) have been seen venturing close to homes where they are "most likely to find insects due to the warmth" [*53]; 7) "A trash-based ecosystem"[*54]; 8) "Over 17,000 species threatened by extinction" [*55]; 9) "Corbett has maximum number of tigers" [*56]; 10) A wounded German shepherd mix "intelligent and calm," a puncture hole in his leg, blood running from his nose and paw, knew precisely where to go: the emergency room of a hospital at the San Juan Regional Medical Center in Farmington, New Mexico. [*57]

"Statistical Incongruities"

Several years ago we were invited to visit a research area in the first national park of Brunei, Ulu Temburong, and were deeply inspired by the fact a relatively wealthy nation of 400,000 people had set aside in perpetuity a stunning and biologically critical portion of Borneo for this and all future generations to marvel at; not to exploit, but to live and let live. Through multiple lightning storms, fantastic downpours, exhilarating aftermaths teeming with the symphonic paens to all the meteorological excitement by scores of birds and primates, we sat in the mud high on a steep pulse of forest, marveling at this remarkable sanctuary.

Of course, one could argue that with a largely oil-derived GDP of over $50,000, such a country can well afford to set aside 56.2 percent of its total terrestrial extent into conservation areas, versus a world average of just over 10 percent. Containing some 6,000 species of flowering plants and 157 known mammals, Brunei's quasi-paradise of a country is, nonetheless, not immune to all those plights making for endangered species.

The nation has at least 124 species in a threatened category, in addition to a known 500 parrots or so legally traded out of Brunei annually. [*58] For comparison sake, in the U.K. 10.5% of the estate is protected, roughly one/fifth the percentage as Brunei. Of the U.K.'s approximately 1623 known flowering plant species, fifty mammals and 229 breeding bird species (246 avifauna in total), 52 of those bird species are red-listed by the IUCN and duly acknowledged by the British Trust for Ornithology. [*59] Other data suggests a total in Britain of only 27 threatened species altogether. [*60] Such discrepancies mirror diverse monitoring techniques, measures and criteria, perception, standard deviation levels, and cultural, religious and political norms that assuredly predispose the outcome of research, one reason current global estimates for the totality of avian species ranges from as low as 9,500 to upwards of 10,000. Such taxonomic relativity also accounts for the uncertainty as to the number of bird species driven extinct by historic outmigration's across oceanic island groups of the Pacific (750 to 1,800) [*61] and the number that have been saved from extinction by active conservation measures, in modern times: usually figured to be sixteen (versus over 100 or more that have gone extinct as a result of the human species). [*62]

If it were not for the Jain ultimate belief in non-attachment (aparigraha) one might want to think of ahimsa and the practice of ecological behavior as one and the same, but not always; not when the hard choices we all must confront arise repeatedly each day, far away from the cloistered confines of a retreat, or some form of sangha, of lay or clerical assembly. In a temple-like atmosphere of the sangha, considerations are generically empowered by religious sobriety, faith, prayer. [*63] But the choices on the battlefields of everyday statistical measurement are usually tinged in some manner with the often overpowering and heartbreaking challenges of attitudes that drive expediencies, however disguised they may be, amid a welter of imperatives. The many incremental compromises chewing away at all quadrants of the global environment stem from inherently biased data, confirming all the gray areas equivalent to compromise.

The life history of the burro in North America is a case in a point, with the U.S. Government agencies directed towards wild land and species conservation, along with countless ranchers and farmers allied in their outright dismissal of this sublime species – *Equus asinus*, reduced today in the U.S. to "31 Herd Management Areas (HMAs)," under the prevailing assumption (notwithstanding the 1971 Wild Free-Roaming Horses and Burros Act) that "the BLM [Bureau of Land Management] would have the public believe that more than 6 million acres of land can support less than 3,000 wild burros." And hence, as of 2012, there were nearly 7 million BLM acres designated as Herd Areas, with no more than "5,841" individual burros therein, suggestive of extremely dubious criteria by which carrying capacities for this benign grazer have been determined. In the continuing legacy of "heartless policy," as Andrea Lococo politely characterizes it, the great symbol who was "Brighty" of the Grand Canyon (the famed book, by the same name by Marguerite Henry, published in 1953) continues unabated, ignored by most. [*64] Is this the best American environmental law and public interest in other species can do to protect one of this continent's most magnificent and sensitive of large mammals? A continuing cultural indifference, in spite of the fact that "Brighty" was a very real person indeed? One who, as Henry wrote, "Long ago, a lone little burro roamed the high cliffs of the Grand Canyon and touched the hearts of all who knew him: a grizzled old miner, a big-game hunter, even President Teddy Roosevelt…"

"Bambi"

Most of us have grown up hearing the phrase, "a lesser of two evils." This is no solution, but may best mirror the human condition. In fact, no less a heartbreak, according to Felix Salten, enshrouded the world of Bambi (**Lebensgeschichte aus Dem Walde**, first published in Berlin by the house of Verlag Ullstein in 1923, then in the U.S., by Simon & Schuster in 1928 – **Bambi - A Life in the Woods**); the world of those talkative and loving leaves that fell off, separated one from the other, lovers parted brutally and swiftly, with the sudden onset of Winter; of the snow hares and foxes who were killed, and sometimes fed upon as carrion by birds. This deep anthropomorphic immersion on the part of Salten, an Ambassador for his species in this matter of the life of a deer. Whatever communions had existed in the Springtime of Bambi's entry into this world – filled with perdurable joy and bountiful excitement and tenderness, fell away as a certain Him with a third arm, meaning a rifle, inexorably entered those same woods and inscrutably enacted his insane violence for no apparent reason. The book was written in the same Viennese Woods of suburban Vienna where Beethoven composed his Pastoral, or 6th Symphony.

The decade in which Bambi was written, and the message this novel foretells, is instructive. The 1920s in Vienna saw, in addition to Salten, who would succeed the essential founder of the State of Israel, Theodore Herzl, as the editor of the Paris-based Neue Freie Presse, enjoyed a Judaic artistic renaissance. Other Austrian Jewish intellects of enormous stature, like Salten and Herzl, included at that time Sigmund Freud, Elias Canetti, Max Reinhardt, Gustav Mahler, Arnold Schoenberg, Hugo von Hofmannsthal, Martin Buber, Karl Popper, Fred Zinnemann, Otto Preminger and Stefan Zweig, among many others. Herzl's efforts – from the Vatican to Russia, in the hopes of gaining international support for the creation of a safe national haven for the Jews had been rebuffed.

In his own world, the one that would so touch a young Walt Disney many years later, the character of Bambi at the discovery of his mother's death represents one of the 20th century's most savage realizations, certainly for young people.

In his essay, "The Trouble with Bambi: Walt Disney's Bambi and the American Vision of Nature," [*65] Ralph H. Lutts describes how Disney had wrestled with the dilemma of whether or not to show the graphic killing of Bambi's mother, ultimately deciding against it. "The mother's death happens completely off-screen. Nevertheless," writes Lutts, "its impact is so great that many people will swear that they actually saw her shot. It was an emotionally compelling scene and even one of Disney's daughters chastised him for allowing Bambi's mother to be killed." [*66]

Lutts cites movie analyst Pauline Kael, as well as Stephen King, both of whom admit that the killing of Bambi's mother left a frightening reverberation. Said Kael, "I have never heard children screaming in fear at any of those movies we're always told they should be protected from as they screamed at Bambi and Dumbo. Bambi's mother is murdered, Dumbo's mother is goaded to madness and separated from Dumbo; those movies really hit children where it counts."[*67] And Lutts adds, "Another critic called the death "surely one of the most devastating moments in any movie."

"Who, Then, Do We Consider Sentient, Vulnerable, Emotional and Worthy of Our Respect?"

In seeking that "lesser" (of two evils) humans seem to have lost the ability to predict and act ethically upon their instincts, particularly with regard to our empathizing with all other species. How could it be otherwise, if "less is more" when it comes to our typically cynical sense of the brutality of nature? During the entire Gulf of Mexico crisis, there was an endless lament for the poor oyster fishermen and their livelihoods, but few were speaking up on behalf of the oysters themselves. Are oysters ("true oysters," members of the Ostreidae family), not living, feeling beings, as opposed to simply being edible by humans? The ancient Chinese and Romans consumed oysters, thinking nothing of the fact the oyster feels pain. *Crassostrea virginica*, the keystone oyster that forms the symbiotic basis of the reef and benthic assemblages, or pillars of the entire coastal waters of the Gulf of Mexico, is an animal, a Mollusk, a bivalve (any hard shelled organism with two hinged valves) that derives its name from the richest man in Roman history, Marcus Licinius Crassus (115 BC – 53 BC), patron of Gaius Julius Caesar, suppressor of Sparticus, political animal, General under Sulla, whose ambitions would lead him to his death in Syria – his name meaning steadfast, solid, dumb. Yet, to repeat, as Linnaeus understood the creature, a being upon whom all else depended, billions of organisms, from red-orange squat lobsters, red grouper and red snapper, amberjack and triggerfish, to worms, shrimp and blue and spiny crabs; it has eggs and sperm, its very own feet (called pediveligers) and asks only that it be left alone to do what it loves to do, and does best. [*68]

The oyster filters water, gives shelter to countless others, feeds off the mechanism of sunlight and phytoplankton, and is crucial to the spawning miracles of oceanic productivity, not to mention the countless birds whose diets incorporate much that results from this web of marine life. What do we know of oysters? Many think of them has half-animal, half-plant, an acceptable food choice for vegetarians. Pearl divers, and those who dine upon them throughout the world, ask no questions. Who speaks on their behalf? Only those who are angry that an oil spill will kill them, rather than providing fishermen the financial benefits of doing so. Our priorities are only too clear: we care about ourselves, the rest is a luxury. We remain callously embittered and unmoved by the suffering of oysters, and of all creatures in the seas, who cannot change "jobs" whereas we humans can.

Even in the seemingly far removed paradise of southern New Zealand, a recent decision by the authorities has determined that the placement of an oyster farm surrounding a scientific reserve that has been bio-invasive free for six years in its ongoing efforts to stem further tides of extinction in New Zealand – a reserve that is adjacent, no less, to New Zealand's newest National Park - is less important than the cultivation and killing of oysters. So much for that allegedly "clean green" paradise eco-tourists have long confused with New Zealand.

Environmentalism has become the brunt of old criticism, the domain of the elite; a realm of romantic fantasy with far less gravity than that of paying off mortgages on fishing trawlers. And then it seemed to be over, the oil having evaporated somehow miraculously, and in one big poof the orgy of news media came to closure. Not that too many people actually asked the remaining oysters, or the Gulf mangroves what their views on the whole disaster were. That would appear as outlandish as asking Felix Salten.

Against this ambiguous backdrop, what comprises true ecological integrity? When is green truly green? What is the ideal, the practical, the combined wisdom that serves both sectors? Can ethics play a viable role in saving species, or is that merely an economic engine of choice, distributed amongst the demographics wherein priorities are predicated upon human income generation? Are we the first and only solipsistic species in biological history? Who may cast the first stone, in other words? There are shades of green, just as every human being marches to his/her own subjective drum. What changes in people? Do we learn from history? Are biological lessons relevant today, or which ones are, which ones are not? Are ethics, compassion and tolerance traits that can be taught? Passed down from generation to generation? Are they biologically hardwired in our genes? Or have we hit the tipping point that annuls the very possibility of ethics in a species that cares only about its own obstinate survival? Where incremental compromises, reflex hatreds and mythic fears, emboldened selfishness calculated to derive maximum pleasure for human individuals has undermined all presidential or egalitarian behavior on the part of the Earth?

All of these troubling questions remain dangling from that first moment of introspection, whenever it occurred.

"The Genesis of Triage"

Triage is a word conservationists try to avoid. But when English naturalist Chris Packham came forward and questioned whether it was not time to let the Giant Pandas die out as a species with dignity, there was a public outcry. Packham had rightly highlighted the fact Pandas were finding it impossible to adapt to new food sources, other than their highly endangered bamboo species. "Wouldn't you adapt your digestive system to another sort of bamboo so you could stay alive?" writes Kerre Woodham. [*69]

This seemingly simple suggestion is obviously not so simple. And because pandas are so "cute" governments are spending a lot of money per individual to keep them going as opposed to spending that money on far less charismatic creatures, like sea slugs, *Nembortha megalocera*, for example, whose tentacles contain highly evolved nerve endings and synapses where chemical transmitters and signal transduction are comparable at many levels to the workings of the human brain, with its hundred billion nerve cells. Is being "cute" the criterion for conservation resource allocations? This is just one type of problem in reconciling the human condition with the condition of the planet's biodiversity. A problem that was no less so for Salten, or Mahavira.

Another such problem is human diet. This is obviously as central a consideration as anything, where most non-sessile organisms are concerned. Elena Kagan, in testimony before a Senate Judiciary Committee as she prepared to become the 112th U.S. Supreme Court Justice under the Obama administration, was stumped when asked by Republican Senator Tom Coburn of Oklahoma whether the Senate could pass a bill mandating that Americans eat only fruits and vegetables? A great question, one that Americans who made it illegal to consume heroin after millennia of its use, or drive their cars without wearing seatbelts – should find most compelling in an age of global warming and increased public awareness of animal cruelty. Instead, this shoe-in for the Court responded, "It sounds like a dumb law."

Kagan was born in New York, where coincidentally Walt Whitman (1819-1892) enshrined his **Leaves of Grass**. Whitman moved nearby to Camden New Jersey, where his self-published poetic epic would be released in May of 1855. If there is any significance to this connection it is this: Whitman celebrated nature, as did the indigenous Unami, Munsee and Unalachtigo Lenape tribes of the region, or Lenni Lanape. Whitman was no vegetarian any more than the Lanape peoples were. They feasted on plentiful vegetables, to be sure; but also routinely slaughtered at least a dozen species of furry mammals as well as any number of fish species. For Whitman, whose poem is thought to be one of the great ecological revelations of the 19th century, "Song of Myself" (one of the many poems within the large body of work) celebrates Whitman's own acknowledgement of contradictions. Perhaps Kagan was merely following the leader of the Commander in Chief who, days before her testimony, concluded a $1.2 billion poultry deal for trade with Russia, having a lunch with the Russian president that consisted of cheeseburgers and fries. During his campaign, at one point, Obama had declared that there were no vegetarians in his camp (though this was not entirely true as there was one young Jain who served on the transition committee for the President-elect).

The point is this: history, geography, ethics and circumstance are not at fault. Human choices are everything, and triage appears to be a sought after ethic of choice.

Consider the unambiguous declaration in a Sunday New York Times article entitled, and scrawled in blood, "Against Meat – Or at least 99 percent of it." There, the author, Jonathan Safran Foer stated: "According to reports by the Food and Agriculture Organization of the U.N. and others, factory farming has made animal agriculture the No. 1 contributor to global warming (it is significantly more destructive than transportation alone), and one of the Top 2 or 3 causes of all of the most serious environmental problems, both global and local: air and water pollution, deforestation, loss of biodiversity...." "Eating factory-farmed animals – which is to say virtually every piece of meat sold in supermarkets and prepared in restaurants – is almost certainly the single worst thing that humans do to the environment. Every factory-farmed animal is, as a practice, treated in ways that would be illegal if it were a dog or a cat." [*70]

In the same month as this aforementioned report, The New York Times ran another cover on its Sunday edition, entitled "The Burger That Shattered Her Life – Trail of E. Coli Shows Flaws in Ground Beef Inspection System." [*71]

We say "stop for pedestrians" and we see signs for deer crossings and in some parts of the U.S., Oklahoma, for example - bison crossings. But we do not stop, nor flinch – the vast majority of people – when it comes to consuming deer, or even bison. Yet, biologists continue to drive home the point of a real Garden of Eden continuing in those few places on Earth, many just recently discovered, where there is amazing biological camaraderie. In the unexplored three-kilometer wide crater of Mount Bosavi, more than 40 new species were discovered by a team of scientists from Papua, New Guinea, the U.S. and Great Britain. In a country where rainforest is being lost at a rate of about 3.5% each year, the explorers found 16 new species of frogs, three fish, a bat, and possibly the largest rat in the world: all new to science. One of the team members found a new species of silky cuscus, a tree dwelling, herbivorous marsupial to be so innocent, that it came and sat on his shoulder. The volcano last blew some 200,000 years ago, and no human or large predator has been in there, until now. [*72]

"Post Apocalyptic Bioinformatics"

Throughout 2008 in just one region of Southeast Asia, the Mekong River, scientists found on average three new species per week. [*73] Recently in Australia, researchers have uncovered 850 new species in subterranean environments. "What we've found," said one of the scientists, Andy Austin, "is that you don't have to go searching in the depths of the ocean to discover new species of invertebrate animals – you just have to look in your own backyard." [*74]

Such discoveries are, of course, heartening - much like the rebound of the Yellow-Eared Colombia Parrot (*Ognorhynchus icterotis*) from 81 to a thousand in number, thanks to the collaboration of nearly 50 organizations, the acquisition of 10,000 acres of parrot corridor with its critical wax palm habitat and funding from three conservation organizations in particular, ProAves Fundacion, American Bird Conservancy and Loro Parque; or the European Commission's promotion of a 20-20-20 CO_2 emissions policy; California's decision to ensure that breeding hens in manufacturing not be forced to live in "crammed" conditions (said law to take effect by 2015, though the legislation mandating it, Proposition Two, was unanimously embraced by Californians in 2008); a former poet Laureate of the United States, W. S. Merwin, who for decades has been warning of the war humans have been waging against the earth, not least of which, whales.

Notwithstanding the occasional discovery of new species, laws of uncertainty set to take effect sometime in the future, the suffering continues, underscoring by vivid testimony the vulnerability that governs the safe habitat of species so easily caught up in the triage maelstroms into which our one species has placed them all, whether urban, suburban, or wild.

And triage, itself – from the French, trier, to "separate out," "to pick or to cull" in fact has become a sovereign misnomer: we cull everything, without limits. This was not the case, for example amongst Native Americans, as most eloquently articulated by the intoxicating stanza, said to be the last words, of Crowfoot (c. 1821-1830 – 1890) also known as Isapo-Muxika or, in Blackfoot, Issapóómahksika, "Crow-big-foot," chief of the Siksika First Nation in what now Canada: "What is Life? It is the flash of a firefly in the night. It is the breath of the buffalo in the winter time. It is the little shadow which runs across the grass and loses itself in the Sunset." [*75]

When in 1970, the year of the first Earth Day, Paulien Hogeweg and Ben Hesper defined bioinformatics to encompass the study of processes within biological systems driven by information, and abetted by computational biology – the processing of large convergences of data – we wonder if times were not slightly more innocent. For example, there was no Web or Google Earth, and limited public access to much of the proprietary LANDSAT data. Even GPS was still controlled by the militaries of Cold War nations. Extinction rates were scarcely understood. Nor was climate change anything more than a theoretical worry, guided by ancient examples that seemed far enough removed so as to induce little if any consternation.

But now, it is clear that *everything we do matter*s, and enters into the bioinformatic nexus: whether the hundreds of millions of tons of annual CO_2 emissions resulting from air conditioning, a "war" in Afghanistan, American investments in Wall Street corporations that have been largely perpetuating the status quo (depleting ground water aquifers, paving over rivers and wetlands, touring, urbanizing and developing to death the last remaining critical habitat areas of the planet, killing animals for human consumption) and so forth. We seem locked in a stranglehold that is ourselves. The International Criminal Court ("ICC") in the Hague, with its 111 nation-state members, cannot even come up with an agreed upon definition of "aggression" (although linguistic consensus was reached with regard to the shared understanding of "genocide, crimes against humanity and war crimes"). Only one of the member countries of the ICC, Great Britain, has fully agreed to adopt the provisions and obligations of the 1998 ICC Rome Statute, though 60 countries ratified its principles as of July 1 2002.

"Ecological Culpability"

Such "Bambi" and oyster dichotomies, among all the transparently non-triage based killing, begs the issue of our moral and irredeemable ecological culpability as a species; our sanity; the very strangeness that anyone who has ever wandered through a great museum, like the Metropolitan Museum of Art in Manhattan, New York must wonder, gazing in particular upon the dominant colors and themes and magnificent innocence, melancholy, splendor and tragedy that collectively evince humanity's dual capabilities, creation and destruction. It is a crisis that elects every religion, tradition, literary and cultural fragment. Such paradox involves all geography, all life forms, as we continue to elevate ourselves onto the summit of this mountain we think we have conquered, like the famed Naram-Sin stele (Sin being the operative word). This Akkadian moon god who ruled with an iron clad (2254-2218), the grandson and successor of King Sargon of Akka, is seen in the small relief trampling the enemy his army has conquered, ironically – the Lullibi people – and Sin – standing twice the height of his own spear carrying soldiers seems ready to conquer the mountain that juts conically before him. Discovered partly cracked in the lower Zagros Mountains at Susa in 1898, this ultimate commentary on human hubris and wrath is housed at the Louvre.

Even earlier artistically rendered windows on human aggression can be found in some of the Namibian Mesolithic petroglyphs in which are figured men throwing speers at one another dating 5,000 years ago.

And yet, taken together, the good and the hideous, who doesn't want their loved ones to share in the bounty of beauty and joy that is this world? To partake of the sublime vastness of the creation with its near infinity of beings, ecosystems, landscapes, the ever sentient co-habitants who are equally involved with each sunrise and sunset, storm and crashing waves, ten-hundred-billion little humble tunes, nuances, hints, intimations, glints, and motes that give us a sense of grace and satisfaction as members of the vast biological community on Earth? It is that collective of life, spanning trillions of individuals and uncountable generations that requires the courage and power of human conscience to uphold, given our ungainly role in the planetary jigsaw puzzle; a recognition by humans that there is a sublime and fragile character to the world that is largely within our awkward power to sustain or undermine, according to the first introspective epiphanies, as well as the modern paradox, most eloquently underscored in the famed essay by Garret Hardin, "The Tragedy of the Commons," published in Science, December 13, 1968. The hugely influential essay was first delivered as a presidential address before the Pacific Division of the American Association for the Advancement of Science at Utah State University, Logan, Utah, 25 June 1968. [*76]

Charles Darwin made much of the three endemic mockingbirds on Espanola Island in the Galapagos in 1835, but was in no position to draw ethical conclusions from their vulnerability, much less any "imperative" that might assuage today's rapacious destruction of wildlife worldwide, in the guise of human consumption or ecological stupidity.

"Emily Dickinson's Birds"

There is no light at the end of such tunnels other than that vague clamor in our hearts called hope, what poet Emily Dickinson (1830-1886) referred to as "the thing with feathers." That feather, and the hope it has always connoted, contrasts with "a Funeral in My Brain," another one of her poems, drawing attention to her ineluctable solitude, and to the state of affairs that would see the last 22 years of her brief 56 years of life swamped by illness (eye and kidney problems) and the deaths of her father, her mother, several friends, and her nephew. The three existing Dickinson collections, a total of 1800 poems, many written in pencil, without titles, were gathered after her death by her sister Lavinia. During her lifetime, Dickinson published only seven poems. Moreover, it was not until 1955 that the first complete set of the original poems were published together following the rediscovery of the original works by Thomas Johnson.

Perhaps what stands out most poignantly in Dickinson's three-stanza poem about hope as a bird are the last two lines, "Yet, never, in extremity/It asked a crumb of me." It is this "crumb" that strikes at the heart, both as metaphor and reality. Crumb is one of the stranger words in etymological history, with roots dating back to Attic (ancient Greek) becoming common before 1200 A.D., and encompassing German, English, French, Latin, Slovenian, Greek, Danish and Russian. In fact, the word seems to track with the migration of birds from the Arctic to the Mediterranean, which raises a host of contemporary provocations. [*77]

For example, the island of Cyprus is mid-way along the annual migratory route for millions of birds from Europe to North Africa, 90% of whom are endangered to various degrees. Yet, in the year 2008, poachers on Cyprus killed "over one million songbirds" that had stopped to rest, seek food and water, during their migration. [*78]

The poachers killed them most likely for "culinary delicacies." Such leaps – from a secluded poet buried at the West Cemetery in Amherst, Massachusetts – to the Mediterranean, are not rare exceptions. Songbirds have migrated from North America to Brazil for millions of years. In fact, in the 1850s, Henry David Thoreau was the first to make notes of the arriving songbirds to Concord, Massachusetts. Two later ornithologists chimed in with data extending well into the 20th century, William Brewster and

Ludlow Griscom. Then, in 2008, the same year BirdLife Cyprus reported its statistics on poaching of songbirds, a study came out from Boston University that showed at least 24 species of songbirds arriving earlier in Concord than during Thoreau's time. [*79]

What is important to realize in drawing such associations is the fact that songbirds (all vocally gymnastic members of the Passeriformes order), comprise some 4000 species, or nearly 40% of all known bird species in the world. To discover such attrition on one small island, is horrendous, but this is the state of the world now: global warming, poaching, utter disarray. Yet hope, which has also been thought of as the most logical conclusion to an illogical world, is our only common tool for jumpstarting some form of ecological redemption. Hence, Emily Dickinson's extraordinary intuition and importance to the realms of ecological literacy, a phrase easily accessed in terms of many of its component parts and challenges by a singular review of the U.S. Department of Interior's "Facts," for example. Countless documents are accessible for the DOI, obviously, but what should arouse any student's interest is the well-coordinated corpus of complex responsibilities as well as the inevitable compromises. For example, in delaying the listing of the Greater Sage Grouse as endangered (a flagship [indicator] species *Centrocercus urophasianus*, that has already lost at least half of its critical sagebrush habitat, with more loss projected). Former Interior Ken Salazar declared "We must find common–sense ways of protecting, restoring and reconnecting the Western lands that are most important to the species' survival, while responsibly developing much-needed energy resources." [*80] As can be expected, energy companies, including wind, oil and gas, were delighted with the ruling that the grouse would not be listed. The complexities are worth exploring in some depth because they rightly point the way, in each instance, to the same need for hope that must translate into protections.

Just as bioinformatics has morphed into the nanoseconds and exobytes of an information avalanche, so too, metaphysics is a fundamental set of possibilities more harried, stretched and at real risk than ever in our history.

In the case of avifauna, as of October 2009, a known 1217 bird species were "deemed endangered or vulnerable to extinction." [*81]

Migrations of Caribou in Alaska and the Yukon, of Elk in Wyoming and Montana, Wildebeest in East Africa and the longest of all known mammalian migrations - those of the Gray whales who travel up to 12,000 miles (roundtrip) from Baja California to the Arctic and back, feeding on increasingly-scarce amphipods on the ocean floors – all require places to rest along the way. Marine no-kill sanctuaries. Not unlike those Medieval English Churches providing asylum. A Ruby-throated hummingbird weighing less than 5 grams requires weeks of nectar binging to acquire the necessary fat storehouses to fuel its 24-hour non-stop flight across the 600 miles of open waters between Florida and Mexico.

Some albatross and terns can fly for thousands of miles without stopping (in fact, the Arctic tern has been known to fly 18,600 miles annually, stopping only to catch fish along the way in flight), but when they do stop to breed, they are exhausted, as one would expect, and they require safe havens. To store up to 20% of their body weight as fat, birds become what is known as hyperphagic prior to their migrations, eating as much as possible, whether insects or the aforementioned hummingbirds; nutritionally fitted by whatever other food supplies they depend upon, in a primordial and knowing anticipation of the metabolic output that will be required for their upcoming journeys, much like an amphetamine-induced hyperactivity. Along the way, many will stop to rest, it they have that luxury. Some do not.

Songbirds rest during the day and keep their altitudes to below 2000 feet, typically, whereas Bar-headed Geese have been known to fly directly above Mount Everest, at over 29,000 feet. Songbirds fly at night. Those "stopover sites, or staging areas" as they are called are critical rejuvenation areas. Birds may spend weeks there, or just a day or two, depending on numerous components. But what has recently emerged as an important factor in the success of these migrations are the increasingly scarce habitat areas suitable for their needs, that may be as small as 11,000 square feet, or about a quarter of an acre. Thus, backyard refugia have recently emerged in the scientific literature as life-saving spots across the planet. [*82]

In a study of plant diversity, density and distribution within one of the most northerly of all neotropical rainforest parks, the Sierra de los Tuxtlas, in Los Tuxtlas of Veracruz state, Mexico, scientists discovered that even according to a "standardized sample size" of "0.1 ha" or slightly less than a quarter acre, plant diversity can flourish, and similar findings for birds have been reported. Just northwest of Los Tuxtlas, the Reserva de la Biosfera "El Cielo" - the most northern fringe of the Amazon in Mexico's Tamaulipas state – contains the largest populations of healthy wild jaguars. Locals see them all the time and – like those hundreds of crepuscular fruit bats living beside clusters of coconuts on trees across Mozambique's Isle de Inhaca, in the Mozambique Channel –human population nearly 9,000 – all living in peaceful coexistence.

If a backyard, even potted native plants on a balcony can help in the effort to restore some balance to the ecological tumult our species has precipitated, then the world is indeed watching and listening to our individual efforts. That "if" is a proposition filled with doubt, contention, disagreement, and occasional concord.

The insistent ethical stance that proclaims the defense of the innocent as a central credo must also confront a predominant difficulty, within the Jain context. Because the Self, in the Jain universe is not ultimately or even incipiently disassociated from all others, the dilemma, in a world of so many multitudinous Beings, is at once reductionist in its effect. We look at so many linked crises and pains, not indifferent but ineffectual, by and large, to impede the cascade of cries, or staunch the myriad floes of blood.

Whatever successes we can look to are inextricably shadowed by the frustration of every Sisyphean cliché, one step forward, multiple steps back, that is, when bearing a cross, or feeling the deep empathy across family, community, ethnic and species boundaries.

A "forest man," as he has been characterized in the Northeast of India, has, since 1979 been singlehandedly planting trees at his own expense on the Majuli Island in the Brahmaputra adjoining a vast sandbar that looks like a science fiction mirage. His ethos is clearly sincere and monumental. Thus far, Mr. Jadav Payent has planted an area known as the Molai Forest that is 1,360 acres, or nearly a third larger than New York's Central Park. With those trees have come back a population of Asian elephants, Bengal tigers, even a vulture presumed to be extinct for several decades. That, of course, is the vast step forward in every human being; a clear window on ourselves. But, echoing Sisyphus – a true ecological battle that is terrible lament more than myth – would be the fight to save the Greater Sage Grouse in Wyoming. This largest of all North American grouses (*Centrocercus urophasianus*) is also among the most precocious, mysterious, melodious, gorgeous and soon likely to be critically endangered of all avians in North America. It is an obsolete battle between energy developers, farmers and ranchers and politicians driven by zealous constituents who, at this point in time appear to have little compunction in making this miracle of a bird and all her progeny, simply go away. But, like most bird species who are dependent on specific habitats, there is nowhere else for *urophasianus* to go, other than where it is – patches of habitat where its leks (mating areas) are fragmented across Wyoming, ten other states and two Canadian provinces. In nearly every case, modernity is eroding the prospects for this bird.

Each of these individual biographies in peril –the Greater Sage Grouse, a forest in India - is like an Emily Dickinson bird, in need of crumbs, desperate for our help, however ungainly we may be in our – at best – pleading stewardship.

The nature/nurture socio-biological debates given to questioning whether our genomes and/or emotional predispositions are generous and unstinting or selfish and exclusive breaks down into even more difficulties when we place the one beside the many; perceive the individual within the collective, which is precisely what Emily Dickinson accomplished with her poetry that referenced, or focused upon birds "222 times." [*83]

"When Numbers Challenge Even Our Most Beloved Philosophies"

Dickinson's single bird, that bird's soul, registers in an infinity of calls to love. In one bird's future are all the stakes laid clearly before us; reiterated in so many pieces of the biological ramparts as to challenge, if not entirely overwhelm our ability to love successfully and unconditionally.

At the root of that dilemma exists the notion of God.

At the heart of that fallacy in idealism springs the clichés of every hope eternal.

To attempt to quantify quality and qualify quantity drives mad the most pellucid of intentions. But there it is – ecological and emotional conundrums forever locked at some level of friction -regardless of our cognitive frustrations and prayers of reconciliation.

Consider marine invertebrates like krill whose numbers comprise hundreds of millions of tons, yet, inside those numbers are trillions of individuals. Avians, marine mammals and squid depend on krill for their survival, as far as can be determined. This food chain does not speak of ethics. But then, we are trapped in an additional cul de sac of numbers connoting life forms about whose abstract mathematical bubbles and interpenetrating parallel universes we remain all but locked out, mute, uninvolved in the unimaginable quantum of life, namely, as many as 180 trillion beings, if one includes the following: 44 different bacterial species on a single human forearm; 600 species of bacteria in the mouth, more still in a human colon; ten trillion cells per human body; seven million follicle mites in our eyelashes, dominated by two *Demodex* species, the smallest known arthropods – invertebrates with jointed limbs and segmented bodies which include among their rank, of course, crustaceans, insects, spiders and centipedes. How many *Demodex* thriving – more worlds within worlds -in the fur of a grizzly bear? How many bacteria in the gut of a wolverine, or Blue Whale? How many symbiodinia (massively numerous endosymbiotic dinoflagellates) within any surviving coral reef, or other biological iterations throughout the vastness of a single redwood tree?

Quantum mechanics and Ultimate Ensemble theories, applied principally to dimensions, exponential quanta like stars and star systems, when set to work in grasping the biological string theories of Earth conclude without the least readable type... grasping at grains of sand that have, themselves, split into multiple universes at the very moment we look or try to hold on to them. This goes beyond the Anthropic Principle which was at the heart of Bishop Berkeley's intimate contention. These numbers defy mathematics in such a way as to make it clear that numbers, however multitudinous our computations, visualization modalities and re-arrangements in false or true color, are no substitute for all that life herself: all those lives, or nigodas and jivas, as Jains would cosmologically reference these biological atomic charts.

A team of scientists at the University of Georgia led by microbiologist William Whitman estimated that the total number of bacteria in the world number "five million trillion trillion, (a five with 30 zeroes after it.") This was a 1998 study, "The First-ever estimate of total bacteria on earth..." [*84] Moreover, viruses outnumber bacteria with some "10 million virus particles" in every drop of seawater, "100 million" per "pinch of soil" and a world total of something like "1 with 31 zeroes behind it." [*85]

"The Pain Points"

By 2050 an estimated 120 billion farm animals are likely to be slaughtered annually, with the number fast increasing. [*86] In 44 states in the US it is still legal to eat a dog. In 1991 a Chinese American who'd opened a "live poultry" company in Southern California was told by the local Planning Commission that complaints of "rotten stench emanating from the building" would require them to shut down their company within one year. The owner's response, "I'm hoping the city will do the right thing and see the kind of service we are providing for the community… A slaughterhouse sounds funny, but it's very comforting for a lot of our people. It's a place where they can go and it feels like home." [*87] The numbers of vertebrates such as dogs pale by comparison with total numbers of bacteria and viruses. With something like 40 million bacteria in a single gram of soil, it is probably a truism that by most neurological barometers, the torture of a canine will exert far more horror and compassion in most people than the manipulation, as viewed under an electron microscope, of an individual bacterium.

So while pundits focused on environmental and ethical history can point to any number of wake-up calls, in the end, we find ourselves as a species schizophrenically in stark contrast with much of the natural world. Our bodies are bacterial battlefields, whilst our consumption, every move we make, ignites other violence, from a Southern California slaughterhouse, to a recreational vehicle heading out at sixty miles per hour towards some National Park, killing, in the process, untold numbers of invertebrates, and the occasional bird, or mammal. And at some of those national parks, particularly in Alaska, it is legal to hunt large vertebrates, and birds.

Yet, our capacity for peace and non-violence is without doubt. It is there within us. Think of those magnificently voiced echoes floating down from Portia's confrontation with Shylock in Shakespeare's "The Merchant of Venice," - "The quality of mercy is not strain'd,/It droppeth as the gentle rain from heaven/Upon the place beneath. It is twice blest: It blesseth him that gives and him that takes." (Act 4, Scene 1, lines 180-187.) In Sir Thomas More's **Utopia**, first published in 1516, the killing of animals was deemed so horrible that "freemen were not even allowed to witness it lest their human clemency be eroded." [*88]

"The Birth of Tragedy"

In Spanish philosopher Miguel De Unamuno's classic, **The Tragic Sense of Life**, he quotes French theologian and translator of Dante's **Divine Comedy**, Huges Felicite Robert de Lamennais (1782-1854), "Shall we, losing all hope, shut our eyes and plunge into the voiceless depths of a universal skepticism?... Nature does not allow it; she forces us to believe even when our reason is not convinced. Absolute certainty and absolute doubt are both alike forbidden to us. We hover in a vague mean between these two extremes, as between being and nothing... But it is not given to man to annihilate himself; there is in him something which invincibly resists destruction, I know not what vital faith." [*89] Faith in oneself must, by every Jain injunction, and in the culminating thought of both Aristotle and Kant - necessitate faith in others. In terms of political governance, that ethical relationship between the one and the many was deeply ingrained in Mahavira's legacy. Jain texts are unambiguous with respect to the fact that a muni, or Jain holy person "should not kill any living being." [*90]

It is one thing to feel the compulsion of a beautiful ideal, but quite another to manifest it in this world, much like the definition of genius as one percent brain power, 99% sweat equity. Historians like Arnold Toynbee, Oswald Spengler, Clarence Glacken among others have cited nearly two dozen civilizations, such as the Aztecs, which slipped away into extinction. This is but one of many reasons former German Pope Benedict XVI, (Joseph Ratzinger's) third trip to the site of the Nazi concentrations camps Auschwitz and Birkenau, where approximately one-and-a-half million Jews perished, one-fourth of the six million or more Jews massacred by the Nazis, was so provocative. He said, whilst standing there, "In the end, there can be only a dread silence, a silence which itself is a heartfelt cry to God: Why, Lord, did you remain silent? How could you tolerate this?"

In writing about **Don Quixote**, philosopher Georg Lukacs described it as "the epic of a world that has been abandoned by God," adding that "...meaning can never quite penetrate reality, but that without meaning, reality would disintegrate into the nothingness of inessentiality." [*91]

"The Legacy of Henry III"

In England as early as the 12th century Henry III named Waltham Forest a Royal Preserve, later to be known as Epping Forest, 6,100 acres of mixed biomes, the first politically protected area in the world. If "tragedy" had wrangled humanity into its unflinching corrals, Epping constituted the first bona fide attempt at redemption. And because it expressly equated with conservation, as we think of it today, not merely a special domain for royalty with all its hunting and lumber rights prohibited those with "lesser" power in English society, we can assess Epping as a global base-line for ecological protective measures within metropolitan statistical areas, whether Rio de Janeiro's Tijuca National Park or Edmonton Alberta's River Valley, a parkland 22 times larger than New York City's Central Park.

Today, some 66% of Epping is designated as a Site of Special Scientific Interest and a Special Area of Conservation. In the heart of this protected area sits a house known as the Great Standing, constructed for Henry VIII in 1543 as a kind of platform enabling royal guests to watch and/or participate in hunts with crossbows; what the City of London's website describes as "a venue for royal Tudor 'corporate hospitality' to show off the wealth and power of the king." Henry III was no vegetarian, that much is clear. But, to his credit, he appears to have loved the forest.

More recently the name was changed to Queen Elizabeth's Hunting Lodge, though no hunting is allowed these days in the park. But the original designation underscores the ambiguous status of the first protected area in the world, which by today's standards is not always a given, despite legal statutes. Numerous national parks from Alaska to South Africa allow hunting or culling. The paradise of conservation is a myth, both Western and Eastern, Southern and Northern. Paper parks in Indonesia allowing multinationals to burn down tropical peat forests in order to reap the profits to be made from palm oil plantations, or zircon, are cases in point.

Instances of violence resulting from protection, whether against elephant and rhino poachers in parts of Africa, or the case of the Bishnoi martyrs of Rajasthan's Thar Desert are numerous. In this latter case, one must look to the 15th century Bishnoi saint, Jambaji, who elaborated upon 29 potent ecological principles, and wrote them down in a book that is largely adhered to this day. In 1730, in allegiance to Jambaji, the first tree huggers exerted the power of their collective conscience to protect the sacred local trees known as Khejarli (*Prosopis cineraria*) against poachers. 363 people were martyred doing so. [*92]

The incidents of self immolation and other forms of self-sacrifice in an effort to quell violence – violence against trees, habitat, other animals including people - are legion, from Israelite martyrs at Masada described by Josephus, to that of the Vietnamese Thich Quang Duc from the Lin-Mu Pagoda in 1963, to the death of 51-year old British animal rights activist Barry Horne in 2001 following his lengthy hunger strikes while serving an 18 year sentence for his direct actions to help shut down companies that partook in vivisection, fur and leather production. [*93]

"Beyond Tragedy: The 'Eustachium'"

Today's Epping Forest provides scores of ecological educational opportunities for children each year, with the superb staff of the Field Studies Council at Epping Forest Field Centre engendering such courses as Forest and Pond Explorers, Woodland Navigators, Pond Professors, Creature Couriers, Fabulous Minibeasts, Minibeast Madness, Easter Bunny Trail, Underwater and Woodland Wonders, Terrific Trees, Teddy's Forest Adventure and the like. [*94]

At a time in its early history when hunting was allowed at Epping Forest, a quite singular work of art recalled, without so much as saying so, the world of Mahavira. Painted around 1440, it was called "Vision of Saint Eustace," by its creator, Pisanello, also known as Antonio di Puccio Pisano, or da Cereto. Today the painting rests in London's National Gallery. Pisanello took the nobility of wildlife to supersede the hunting instinct of man. There he is on his horse, out for a day's hunt – this Patron

Saint of Hunters - when a vision of the crucified Christ appears to him among the animals, particularly the stag, the deer's huge brace cast precisely to either side of the aura of the crucified Jesus, the whole image perfectly choreographed so as to make the getting of religion (animal rights) a visual compact, or trust, that would never occur quite so literally again in art history. Giorgio Vasari hailed Pisanello as an incarnation of Phidias and Praxiteles, harkening back to the reality of pantheism in Pre-Socratic Greece, while straying far from the truth of a meat-eating Europe.

Dürer's 1501 etching of this vision "Eustachium," [*95] places the stillness of the revelation in a landscape of hybrid poignancy. The five dogs in the image comprise two breeds of Hungarian (Magyar) hunting hounds: two Agars on the left, three Vizslas on the right. Some of the smaller Agar breeds from Dürer's time have gone extinct, particularly the ones used to hunt hares. [*96]

Pisanello's pictorial conversion –greatly amplified by Dürer - to a non-hunting narrative remains out-of-step with the majority of human beings. This is the most profound of all metaphysical realities: our dualism.

Getting beyond the tragedy of a schizophrenic species will require nothing less than that kind of Dvorak "Serenade For Strings in E Major, Op. 22" that somehow manages a majority out of its minor keys. Put differently, an existential biology willing to gather the forensic evidence of its paradise exemplars and put them forward in something other than mere relics, like those petrified beech trees from the Antarctic washing up on the shores of New Zealand's Catlins, somehow invoking a great sigh before the silent mysteries that have preceded us. Today is our only chance. Not yesteryear.

This is a "new nature," which the Authors have referred to in earlier works. The evidence is clear enough. It includes John Muir's writings, particularly on other species. [*97] There is the example of St. Francis speaking to the birds. [*98] There is, of course, Mahavira; [*99] the glories of the Persian pictorial Renaissance – certainly achieving their height in the paintings of the **Shahnamah-yi Shahi (The King's Book of Kings)** with its 258 lavish, mostly paradise landscape imagery by the brilliant Sultan Muhammad Mosavar Tabrizi of the Safavid Dynasty(1501-1722).

Comparable European landscapes obviously include the florals and paradise scenes of the "Velvet Breughel" (Jan Breughel the Elder, 1568-1625), as well as the many versions depicting "The Entry of the Animals Into Noah's Ark"; those many works detailing the inhabitants of the Garden of Eden by the aforementioned Northern Flemish artist, Roelant Savery (1576-1639) whose "Dodo in a Landscape with Birds and Animals" circa 1629, was painted after a stuffed specimen, years before the last and final sighting of a Dodo in the wild.

But it was Dürer – not in his "St. Eustace" so much as his "The Virgin Among a Multitude of Animals" that most excites the entire legacy of humanity's spiritual ecology. [*100] The symbolism was ripe for the beginning of the 16th century, with even the odd tinge of humor, suggestive of the Absurdist allegories that Cervantes would play with a century later. But Dürer is still dependent upon the commissions and politically-correct approbation of a society unclear as to how familiar it might venture within the orbit of a real God, and of a natural history that was teeming with human discovery, new species, continents, new stars. Columbus, three years from his death, was trapped near the end of his fourth and final journey in St. Ann's Bay, Jamaica. He and his two brothers had already made three voyages in vain, in the sense of having unleashed monsters they could not have understood, as well as the hatred of Governors, Hispaniola locals, even the relative indifference of King Ferdinand, who'd let Columbus languish in jail, in chains, for six weeks.

But in Dürer's testimony to a divine paradise and all that it offered humanity, both symbolically and in everyday promise, we see the commingling of burros, Joseph, the Virgin and Baby Jesus not simply laid out within an aura of destiny, but beside a cornucopia of other species, from a crab to a fox; a dog, a flock of sheep to a parrot. But this was a composition half-a-century prior to the French and German Renaissance in ornithology – with Pierre Belon and Guillaume Rondelet; and long before the Dutch morphologist Volcher Coiter or the great Swiss Konrad Gesner's **Vogelbuch** and **Icones avium omnium** (1557). Columbus, in the Spring of 1503 had sighted the Caymans, but this was as bizarre an event, in terms of its potential meaning to Dürer's Virgin, as one can possibly imagine. Not even a shift in the equation of an archetypal nostalgia for Paradise; an altogether revolutionary change in perspective that should enable all future generations to more thoroughly grasp that life co-existing all around them.

What is glorious and startling, however, is Dürer's perfect vision, laid out in meager but telling watercolors (at Vienna's Albertina) with an emphatic overture to an engraving, to be accomplished by some subsequent artist. As it happened, that artist turned out to be the brilliant son of Jan Sadeler, engraver of the "The Creation of the World," (discussed in Chapter One) namely, his incomparable Aedgidius II (ca. 1570 Antwerp – 1629 Prague). The Guild of St. Luke, the Antwerp Cathedral, the Dutch Revolt and the scattering of half-dozen Sadeler family members from Brussels to Vienna would dominate the aesthetic matrix of the late 16th century, culminating in one Emperor's obsession with a single image: Rudolf II, in Prague, and Dürer's "Landscape with the Virgin and Child," as Aegidius Sadeler (c. 1570-1629) titled it, after Dürer's two titles – "Virgin," and/or "Madonna with a Multitude of Animals." Sadeler had been appointed Imperial Printmaker by the Holy Roman Emperor (Rudolf II) and – like Jan Breughel – would be asked to finish what Dürer had started, namely, an engraving of one of the most memorable images in the history of art. Breughel did it in oil. Sadeler mostly likely employed a burin directly upon copper, as had Dürer. [*101] Joseph speaks with a stork. Jesus holds a bird – said to be premonitory of the Passion. All is song, ethology, interspecies passions that elude capture. There is even a sense of humor in the way the Virgin strains to hold on to the Old Testament without dropping the baby Jesus. A stag beetle challenges the puppy. Half of the allegories throughout the Bible populate every nook and cranny of the vast surrounding landscapes. There is such ecological density in this magnificent work of art as to shelter humanity from itself. It is heartbreakingly perfect. One wants to be transported to within its multi-dimensional ethics and never leave.

And just as a shift had occurred tens-of-thousands of years before, our species' observations of nature becoming an inward-dwelling obsession, so too, in this specific work of 1503, Albrecht Dürer presages what would be thought of as "Post-Modernist" in his vacillations between the symbolic, the everyday, and the Sublime, in Nature. A re-telling of Hubert and Jan Van Eyck's "Ghent Altarpiece," the 12 paneled paradise scene, also with Mary and Jesus, countless animals and plant species, painted exactly 70 years prior to Dürer's masterpiece in a combination of tempera and oil on wood. But where, in the Van Eyck brothers, fidelity was targeted at photorealism, in Dürer, it is more the delectable knot at the cornerstone of our complex, all too human relationship to the wilderness. Both masterworks are steeped in our history, but Dürer rejects the pedagogical brilliance of the Van Eycks, embracing instead a rogue theory of life: something beyond the divine, a truce with other species, right here on Earth. It is an amazing accomplishment.

A year later, in 1504, Dürer engraved his famed "Adam and Eve" – celebrated as the most poignant and realistic depiction ever created, certainly a precursor to Lucas Cranach the Elder's (1472-1553) "Adam and Eve" painted 24 years later and today hanging in the Uffizi in Florence, along with various versions by Cranach of the original couple, at the Norton Simon, the Art Institute of Chicago and particularly, the "Adam and Eve" at the Courtauld Institute of Art Gallery in London, where they are surrounded by animals, painted by Cranach in 1526.

Dürer eventually departed for Italy, where the biological influences mounted, but never surpassed what he had achieved with his Virgin.

Now: If, to this legacy, one adds the Utopian socialism of Karl Marx and Friedrich Engels; the life and works of Robert Owen, Jeremy Bentham, Charles Fourier, the Moravians, Shakers, Anabaptists and German Pietists, then one has come to the 19th century American Hudson River School and Transcendentalists at their height, with painters Albert Bierstadt, Thomas Moran, Frederic Church; Luminist George Inness, and of course Thoreau, whose visionary life was carried forward by men like President Teddy Roosevelt and Aldo Leopold, culminating in the U.S. Fish & Wildlife Service; and the first Wilderness Act signed into American law in 1964 by President Lyndon Johnson.

In viewing this erratic and eclectic history, how broad a scope, how remarkable a latitude, of (Saint) Thomas More's "utopia," (1516) the man (canonized by Pope Pius

XI in 1935 and declared the "heavenly Patron of Statesmen and Politicians by Pope John Paul II in the year 2000) who would be executed as a Catholic martyr for refusal to grant Henry VIII supremacy over the Pope Clement VII. More had merely imagined some other reality for himself and, by implication, for the rest of us. And how fitting, with the hindsight of the passion to get beyond the unthinkably gruesome beheadings and massacres, that the very origins of the Greek word for arcadia should be so muddled. Named after the legendary Arcas, that mythic hero who was the son of Zeus and of the nymph, Callisto. Today, Arkadia is a prefecture that includes a skiing resort and the Greek National Road 7, that runs right through it. Moreover, a coal mine and thermoelectric power station are found in the region, which includes a town called Megalopolis, the same term used by French geographer Jean Gottmann (1915-1994) to describe the heavily populated region between Boston and Washington D.C. ("BosWash"). [*102]

Megalopolis – like Arcadia, California, just east of Pasadena, in the 4,000 square kilometer southern California basin that is the freeway smoldering, polluted, drought and fire, crime and endangered species-ridden Los Angeles, in a state headed towards more than 60 million people -is certainly out of step with that wondrous spate of true "arcadias," such as those by Sannazaro (1503/1504) and all the enchanted forests of Bohemia from Diana of Jorge de Montemayor, to Sir Philip Sidney's **The Countless of Pembroke's Arcadia**, first published in 1590, four years after Sidney's death.

From Virgil's ancient **Eclogues** to Nicolas Poussin's famed "Et In Arcadia Ego" ("Even in Arcadia There Am I" – painted in 1637) and Thomas Cole's noble version of "The Arcadian," or "Pastoral State" on the Upper Hudson, painted in 1834, this theme has – despite the irrefutable geographical contradictions – infused its lyrical utopias into our counter-utopian behavior.

In the same way that Dürer had thoroughly portrayed for us the sickness and diabolical nature of a species that celebrates its crucifixion. [*103]

Ultimately, this same contradictory nature of conservation, encompassing the history of paradise, utopia, of non-violence, and all of its shortcomings and vast idealism yet to breathe, can be discerned with enormous complexity and clarity in the works of John Ruskin, particularly his great, little book, **Until This Last**, the book that transformed Mahatma Gandhi's thinking. [*104]

No more than 12 percent of the terrestrial Earth remains protected to any degree, the oceans far less than 2%. [*105] And we know only too well that "protection" does not connote "no kill," let alone "no pollution". We are an impossibly conflicted species, with a hypothetical set of logical outcomes that range from sheer insanity and collective suicide, to survival, to survival with dignity. The pace of ecological destruction easily dashes any individual's grasp, by enormous margins. Yet, psychologists schooled in decades of biophilia and fascinating statistics grapple with the hope, however distant, that our familiarity with nature might be the only lasting medicinal for all concerned. [*106]

We have spent our time in Nature, for however many eons – 50 million years or more, if one includes all our diverse and globally diffused hominidae ancestors – from present day Ethiopia, Kenya, Tanzania and Mozambique, to Montana - whose genes together brought us to consciousness. Yet, that represents very little time, against the entire span of organic life forms on Earth. Our background check encompasses roughly 1.5% of the biological record. In other words, we were born not yesterday but today, just a few minutes ago. Take that as a head start towards active compassion.

If there is anything positive to glean about a newborn; information that can help Others to survive, then it is clear we have no other choice but to nurture that infant, and rapidly digest the best of that which we are, throwing out that which we have proved does not work; embracing the heart, embracing life, all life, lest we simply disappear, unaware of the chain of events, all the other countless lives squandered, and the astounding possibilities we still have time to honor and – side by side, in harmony and out of harm's way –live to the fullest. This ethical disposition is our only chance. Our today and our tomorrow.

Footnotes:

*1 **Medieval Sourcebook: Salimbene – On Freederick II**, 13th Century, **From the Chronicle of Salimbene**, thirteenth-century Italian Franciscan, as translated and paraphrased by G. Coulton

*2 http://www.abdn.ac.uk/stalbanspsalter/english/commentary/page056.shtml; See https://www.abdn.ac.uk/stalbanspsalter/english/essays/introduction.shtml

*3 Semiotica 42-1 (1982) 25-82. 0037-1998/82/0042-0025, © Mouton Publishers. See the "International Conference: Zoosemiotics & Animals Representations," Tartu, Estonia, 4-8 April 2011, http://www.ut.ee/SOSE/conference/2011_zoosemiotics/

*4 http://en.wikipedia.org/wiki/April_Fools%27_Day - cite_ref-6"Chaucer in the Twenty-First Century: The Nun's Priest's Tale". University of Maine at Machias. 2007-09-21. Retrieved 2013-04-04.

*5 Argue, D.; Donlon, D., Groves, C. and Wright, R. (October 2006). "Homo floresiensis: Microcephalic, pygmoid, Australopithecus, or Homo?" Journal of Human Evolution 51 (4): 360– 374. doi: 10.1016/ j.jhevol. 2006.04.013. PMID 16919706. This data comports with the newly discovered 35,400 year-old cave painting of a pig deer on Sulawesi.

*6 www.jstor.org/stable/1400182

*7 See the one-hour PBS documentary, "Ahimsa: NonViolence," Written and Directed by Michael Tobias, KRMA/Denver/PBS, 1987; Newly released by Dancing Star Foundation, www.dancingstarbooksfilms.org/ahimsa-dvd/

*8 **Panorama Of Jain Art**, by C. Sivaramamurti, The Times of India, New Delhi, 1983, p.158.

*9 **"The Plant That Could Erase Chronic Pain: An ancient Chinese re**medy shows promise," By Jessica Chia, http://www.prevention.com/health/health-concerns/chinese-remedy-corydalis-pain

*10 p.12, **Tercentenary of `Don Quixote' - Cervantes in England**, by James Fitzmaurice-Kelly, London, Published for the British Academy, Read January 25th, 1905

*11 http://quixote.mse.jhu.edu/Translation.html

*12 http://www.csdl.tamu.edu:8080/dqiDisplayInterface/displayMidImage.jsp?edition=1&image=1620-London-Blount-Addenda-01-001-t.jpg

*13 Cushing Memorial Library and Archives, Texas A&M University, http://www.csdl.tamu.edu:8080/dqiDisplayInterface/doSearchEditions.jsp?ftMode=phrase

*14 The most beautiful of all Munchausen editions probably to be found within **The Travels and Surprising Adventures of Baron Munchausen. Illustrated with Twenty-Three Curious Engravings, from The Baron's Own Designs, and Five Woodcuts**, Cruikshank, George, illustrator. Originally written by Rudolf Erich Raspe, Published by William Tegg, London, 1869.

*15 http://www.openculture.com/2013/12/gustave-dores-definitive-engravings-of-don-quixote.html

*16 See **Maps of Paradise**, by Alessandro Scafi, The University of Chicago Press, Chicago and London, 2013.

*17 See http://rvwsociety.com/i-frame/lark_doc.html; See also, www.rvwsociety.com/journallist.html

*18 "Rapid Acidification of the Ocean During the Paleocene-Eocene," Thermal Maximum," James C. Zachos, Ursula Ro hl, Stephen A. Schellenberg, Appy Sluijs, David A. Hodell, Daniel C. Kelly, Ellen Thomas, Micah Nicolo, Isabella Raffi, Lucas J. Lourens, Heather McCarren, Dick Kroon, www.sciencemag.org Science Vol. 308 10 June 2005, pp.1611-1615, 21 December 2004; accepted 18 April 2005 10.1126/science.1109004.

*19 "Stop Stimulating Demand: Discussion of Ivory Trade Mechanism May Itself Spur Consumer Demand & Poaching," Agenda Item 36 at CoP16 THE determination by the ...eia-international.org/.../Ivory-trade-...CITES (CoP 16) March 2013: Ivory-trade-mechanism-briefing-FINAL-pdf

*20 See **The Guns of August**, by Barbara W. Tuchman, Foreword by Robert K. Massie, Presidio Press, CA, 2004. See also, The Great War: A Combat History of the First World War by Peter Hart, Oxford University Press, New York, 2013. See also, **The Great War: July 1, 1916: The First Day of the Battle of the Somme** by Joe Sacco and Adam Hochschild, W. W. Norton & Company, New York, 2013.

*21 Lecture on Kant's "Perpetual Peace" by Ian Johnston, June 1999, http://records.viu.ca/~johnstoi/introser/kant.htm

*22 **Kant's Critical Philosophy For English Readers**, by John P. Mahaffy, D.D., and John H. Bernard, B.D., Vol. II., **The Prolegomena Translated, With Notes and Appendices**, London, Macmillan And Co., 1889, p.163.

23 "Grizzly federal protection restored," by Carol J. Williams and Julie Cart, Los Angeles Times, November 23, 2011, p. A16.

*24 See "While The World Moves On, U.S. Still Caught In Its Traps," AWI Quarterly, Fall 2013, Vol. 62, Number 4, pp.20-23. n.a.

*25 ibid., p.20.

*26 See "Sprouting a new line of produce seeds," by P.J. Huffstutter, Los Angeles Times, October 20, 2011, pp. B1 and B7.

*27 http://www.washingtonpost.com/blogs/capital-weather-gang/wp/2013/12/17/earth-has-its-warmest-november-on-record/

*28 For Dr. Bekoff's data in **Rewilding Our Hearts: Building Pathways of Compassion and Coexistence**, p. 39 (New World Library, Novato, CA, 2014). For the Faroe Islands data, see, "Wanton Waste And Suffering Suffuse Faroe Islands Drive Hunts," AWI Quarterly, Fall 2013, Vol. 62, Number 4, p. 19, n.a.

*29 **God's Country: The New Zealand Factor**, by Michael Charles Tobias and Jane Gray Morrison, Zorba Press, Introduction by Ms. Ingrid Newkirk, President of PETA, People for the Ethical Treatment of Animals, A Dancing Star Foundation Book, 2011, p.386.

*30 Journal of Agricultural and Environmental Ethics 16, 2003, pp.387-394.

*31 See Gaverick Matheny, "Least Harm: A Defense Of Vegetarianism From Steven Davis's Omnivorous Proposal," Journal of Environmental and Agricultural Ethics, 16, 2003, pp.505-511.

*32 See, in the case of Haiti, http://www.forbes.com/sites/michaeltobias/2013/03/09/how-a-single-national-park-might-help-transform-a-nation-haitis-pic-macaya-2/

*33 http://users.rcn.com/jkimball.ma.ultranet/Biology/Pages/C/Caen.elegans.html

*34 Barnes, Robert D. (1968). Invertebrate Zoology (2nd ed.). W.B. Saunders. OCLC 173898, www.worldcat.org/oclc/173898/editions?editionsView=true...di

*35 See N.H.Putnam, July 2007, Sea anemone genome reveals ancestral eumetazoan gene repertoire and genomic organization." Science 317 (5834): 86-94. Doi:10.1126/science.1139158. ISSN 0036-8075, promo.aaas.org/kn_marketing/biomed/08_Aug07biomed.html

*36 http://www.nytimes.com/2012/12/16/nyregion/gunman-kills-20-children-at-school-in-connecticut-28-dead-in-all.html

*37 JSTOR: Ars Orientalis, Vol. 16 (1986), pp.119-151, http://www.jstor.org/pss/4629345

*38 "Bhakti in the Early Jain Tradition: Understanding Devotional Region in South Asia," John E. Cort, History of Religions, Vol. 42, No. 1, August 2002, pp.59-86.

*39 **Aristotle: Parva Naturalia, A Revised Text With Introduction And Commentary**, by Sir W. David Ross, Oxford, The Clarendon Press, 2001, p. 50.

*40 See Pali Text Society, http://www.accesstoinsight.org/tipitaka/an/index.html.

*41 **A Comparative Study Of Jainism And Buddhism**, by Brahmchari Sital Prashad, Sri Satguru Publications, India, 2nd Edition, Delhi, 1982, p.236.

*42 ibid., p.14.

*43 "A World Of Harmony," by Christopher Knight, Los Angeles Times, November 20, 2011, p. E14.

*44 **"Don Quixote - An Introductory Essay in Psychology,"** by Salvador De Madariaga, translated from the Spanish by the Author and H.M. de Madariaga, The Gregynog Press, Newtown, UK, p. 136.

*45 "A Feathered Nest - Bird diversity linked to increased home values," in "Journal Watch," by David Malakoff, p.12, Conservation Magazine, Winter 2012, and based upon Farmer, M.C., M.C. Wallace and M. Shiroya, 2011, "Bird diversity indicates ecological value in urban home prices." Urban Ecosystems doi:10.1007/s11252-011-0209-0.

*46 For data on children, see Ballouard, J.-M., F. Brischoux and X. Bonnet. 2011. Children prioritize virtual exotic biodiversity over local biodiversity. PloS ONE doi:10.1371/journal.pone.0023152, ibid.,Malakoff, p.10, citing Mark. Z. Jacobson and John E. Ten Hoeve of Stanford University, pp.9-10

*47 "The good companion," Country Life, November 23, 2011, pp.90-91.

*48 AWI Quarterly, Spring 2009, pp.17-18, n.a.

*49 By Libby Bacon and Richard B. Primack, Boston University, "OOS 20-2 –Shake-up in Timing in Ecological Communities: Understanding the Complexity and the Role of Citizen Science," http://eco.confex.com/eco/2008/techprogram/ P10085.HTM

*50 By Jim Tankersley, Los Angeles Times, March 6, 2010, p. A12. See "Energy groups relieved sage grouse won't be listed," by Mead Gruver, March 5, 2010, Yahoo! News, http://news.yahoo.com/s/ap/20100305/ap_on_re_;is/us_sage_grouse/print

*51 October 13, 2009, nzherald.co.nz, www.nzherald.co.nz/animals/news/print.cfm?c_id=500834&objectid=10602796&p...

*52 By Victor Arroyo-Rodriguez, Eduardo Pineda, Federico Escobar, and Julieta Benitez-Malvido, Conservation Biology, Volume 23, No.3, pp. 734-735.

*53 Country Life, January 20, 2010, p.33.

*54 "A trash-based ecosystem – Scientists call it the plastisphere, and it's got them worried," by Louis Sahagun, Los Angeles Times, December 28, 2013, pp.A1, A6.An ecosystem of our own making could pose a threat": "The plastisphere, a marine ecosystem that starts with bacteria on particles of discarded plastic, is drawing increasing attention. Scientists fear it might host pathogens and leach dangerous chemicals."

*55 By Frank Jordans, AP writer, Yahoo! News, http://news.yahoo.com/s/ap/20091103/ap_on_sc/eu_endangered_species/print, November 3, 2010; See also, See "Guidelines for Using the IUCN Red List Categories and Criteria," Version 7.0 (August 2008), Prepared by the Standards and Petitions Working Group of the IUCN SSC Biodiversity Assessments Sub-Committee in August 2008. Http://intranet.iucn.org/webfiles/doc/SSC/RedList/RedListGUidelines.pdf.
See also IUCN -Category 1a Strict Nature Reserve," http:iucn.org/about/work/programmes/pa/pa_products/wcpa_categories/pa_catory1a/

*56 Posted by TigerAngel, http://timesofindia.indatimes.com/Corbett_has_maximum_number_of_tigers/articleshow/2777812.cms

*57 See "Hurt dog limps into New Mexico ER," AP, MSNBC.com, March 16, 2010, www.msnbc.msn.com/id/35896305/ns/health-pet-health/print/1/displaymode/1098/

*58 http://earthtrends.wri.org

*59 http://www.bto.org/volunteer-surveys/birdtrack/bird-recording/birds-conservation-concern

*60 Steadman D (2006). **Extinction and Biogeography in Tropical Pacific Birds**, University of Chicago Press. ISBN 978-0-226-77142-7.

*61 Butchart, S; Stattersfield, A; Collar, N (2006). "How many bird extinctions have we prevented?". Oryx 40 (3): 266–79. doi:10.1017/S0030605306000950.

*62 http://earthtrends.wri.org/country_profiles/index.php?theme=7

*63 See **Jainism: The World of Conquerors**, Volume 1, by Natubhai Shah, Motilal Banarsidass Publishers, Delhi, India, 2004, p.4.

*64 See "Wild Burros –Tough Survivors Face Harsh Landscape and Hostile Management," by Andrea Lococo, AWI Quarterly, Fall 2012 Volume 61 Number 4, pp. 14-17. See also **Brighty of the Grand Canyon,** by Marguerite Henry, Illustrated by Wesley Dennis, Published by Rand McNally & Company, New York and Chicago, 1953.

*65 Forest and Conservation History 36 (October 1992): 160-171.

*66 Lutts, footnote #7. He cites the following references for this: "**Ollie Johnston and Frank Thomas, Walt Disney's Bambi: The Story and the Film** (New York: Stewart, Tabori & Chang, 1990), pp. 170-75. Storyboard sketches of the mother being shot appear on pp. 170- 71; Jamie Portman, "Generations Stunned by Death Scene in 'Bambi'," Boston Globe, 15 July 1988, p. 32; Charles Solomon, **Enchanted Drawings: The History of Animation** (New York: Alfred A. Knopf, 1989), pp. 129-30."

*67 ibid, Lutts' essay, footnote #8. He is citing "Pauline Kael, Going Steady (Boston, Massachusetts: Little, Brown & Co., 1970), p. 225; David Kehr, "'Bambi' Handles Emotion with Maturity, Sensitivity," Chicago Tribune, 15 July 1988, sec. 7, p. L; Stephen King, "A Master of Horror Stories has some Provocative Thoughts about Tots, Terror and TV," TV Guide 29 (13 June 1981): 8.

*68 "Identifying the true oysters (Bivalvia: Ostreidae) with mitochondrial phylogeny and distance-based DNA barcoding." Liu J, Li Q, Kong L, Yu H, Zheng X, Mol Ecol Resour 2011 Sep;11(5):820-30. doi: 10.1111/j.1755-0998.2011.03025.x. Epub 2011 May 19. www.ncbi.nlm.nih.gov/pubmed/215923213.Accessed January 10, 2014.

*69 http://www.nzherald.co.nz/science/news/article.cfm?c_id=82&objectid=10599792

*70 "Against Meat," by Jonathan Safran Foer, The New York Times Magazine, October 7, 2009, www.nytimes.com/2009/10/11foer-t.html?pagewanted=1&_r=1

*71 "The Burger That Shattered Her Life," by Michael Moss, The New York Times, October 4, 2009, pp. A1, A24-25.

*72 See "Lost world of fanged frogs and giant rats discovered in Papua New Guinea," by Robert Booth, The Guardian, September 7, 2009, www.guardian.co.uk/environment/2009/sep/07/discovery-species-papuanew-guinea/...

*73 See "Fanged frog, 162 other new species found in Mekong," by Michael Casey, Yahoo!News, http://news.yahoo.com/s/ap/20090925/ap_on_re_as/as_mekong_new_species/print; See also, "Climate Change Puts New Species at Risk," by Joel Chong, Yahoo! News, at http://news.yahoo.com/s/oneworld/20090925/wl_oneworld/world3669801253908035/print

*74 See "850 Mostly Blind, Pale Creatures Discovered Underground/LiveScience," By Charles Q. Choi, September 28, 2009, www.livescience.com/.../090928-new-species-underground.html; See also "Subterranean spiders among new species discovered under Australian desert," by Louise Gray, Environment Correspondent, Telegraph.co.uk., www.telegraph.co.uk/earthnews/624660/Subterranean-spidersamong-new-species-discovered; See also "Rare Vegetarian Spider Discovered," AOL News, h

AOL News, http://news. aol.com/article/rare-vegetarian-spider-bagheera-kiplingi/714185 - October 12, 2009.

*75 – Quoted in **Sacred Legacy – Edward S. Curtis And The North American Indian**, Photographs by Edward S. Curtis, Edited by Christopher Cardozo, Foreword by N. Scott Momaday, Essays by Christopher Cardozo and Joseph D. Horse Capture, Afterword by Anne Makepeace, Verve Editions, Ltd. and Cardozo Fine Art, 2005, p.29.

*76 See The Garrett Hardin Society, http://www.garrethardinsociety.org/articles/art_tragedy_of_the_comm...

See also "How Many People Have Ever Lived on Earth?" by Carl Haub, PRB –Population Reference Bureau, http://www/prb.org/Articles/2002/HowManyPeopleHaveEverLivedon...

*77 books.google.com/books?isbn=0870239996; See also, www.poetryfoundation.org/poem/171619

*78 www.greenprophet.com/2013/12/cyprus-kills-ambelopoulia-birds/

*79 **Birds of Concord – A Study in Population Trends**, by Ludlow Griscom, Harvard University Press, Cambridge, MA, 1949, pp. 4-47. See also **Thoreau On Birds, Selections from his writings**, compiled and with commentary by Helen Cruickshank, Foreword by Roger Tory Peterson, McGraw-Hill Book Company, New York, 1964.

*80 www.fws.gov/mountain-prairie/species/.../PressReleaseDOI03052010.pdf

*81 www.animalsanctuaryinfo.com/endangered-species-birds-around-the-world/

*82 http://www.birdsource.org/gbbc/science-stories/2013summary; www.allaboutbirds.org/page.aspx?pid=2459

*83 **A Spicing of Birds – Poems By Emily Dickinson**, Illustrations By Early Masters Of Bird Art, Selected and introduced by Jo Miles Schuman and Joanna Bailey Hodgman, Wesleyan University Press, Middletown, Connecticut, 2010, See Introduction, pp. xiii – xxiv.

*84 www.uga.edu/gm/1298/FrontFind.html

*85 See "Virus Myth Busters," by Sharath Srinivasiah, Doctoral Student, and Professor K. Eric Wommack, University of Delaware, www.expeditions.udel.edu/extreme08/microbes/virus-myth.php

*86 op. cit., See Tobias & Morrison's **God's Country: The New Zealand Factor**, pp.66-67. See the most recent data cited by Dr. Marc Bekoff in his new book, **Rewilding Our Hearts: Building Pathways of Compassion and Coexistence**, November, 2014.

*87 "Slaughterhouse faces chopping block," by Corina Knoll, Los Angeles Times, December 29, 2009, A4.

*88 See **Man And The Natural World - A History Of The Modern Sensibility**, by Keith Thomas, Pantheon Books, New York, 1983, pp. 294; Thomas is citing p. 71 of More.

*89 Essai sur l'indifference en matiere de religion, iii partie, ch. lxvii, 1817. Quoted by Unamuno, Tragic Sense of Life, trans. J. E. Crawford Flitch, Dover Publications, Inc., New York, 1954, p.117.

*90 See **Ayaro** [**Acaranga Sutra**], Synod Chief Acharya Tulsi, Commentator & Editor Yuvacharya Mahaprajna, English Version by Muni Mahendra Kumar, Foreword by Dr. Nathmal Tatia, Published under the auspices of Jain Vishva Bharati, Ladnun, Rajathan, India, by R. K. Jain Today & Tomorrow's Printers and Publishers, Delhi, India, 1981, p.277, "Nativaejja kamcanam."

*91 **The Romantic Interpretation of "Don Quixote",** Cambridge University Press, Cambridge, England, 1978, cited in Don Quixote, translated by John Rutherford, Penguin Books, U.K., 2000, p. ix.

*92 www.animalsandsociety.org/assets/library/312_s415.pdf

*93 See www.barryhorne.org/

*94 www.field-studies-council.org/centres/eppingforest.aspx

*95 Albrecht Dürer, "Saint Eustace," B. 57; M., Holl. 60; S.M.S. 32).

*96 See "Discovering the Dürer Cipher : The Dogs of Dürer…By Elizabeth Garner http://art-antiques-design.com/2013/06/16/discovering-the-durer-cipher-the-dogs-of-durer-by-elizabeth-garner/

*97 See **Muir Among the Animals – The Wildlife Writings of John Muir,** Edited by Lisa Mighetto, Sierra Club Books, San Francisco, CA, 1986. See also www.notablebiographies.com/Mo-Ni/Muir-John.html; also www.sirraclub.org/john_muir.../life/muir_biography.html; for one of the best collections of

John Muir materials, see the Holt-Atherton Special Collections: John Muir Papers at www.library.pacific.edu/ha/muir; See also www.johnmuirassociation.org/collection/kimes_collection.ppt

*98 "St. Francis Preaches to the Birds," retold by John Feister, www.americancatholic.org/features/francis/stories.asp

*99 For a definitive life of Mahavira, see the **Kalpa Sutra of Bhadrabahu**, Translated by Hermann Jacobi, 1879. Jaina Sutras, Part 1, Sacred Books of the East, Vol.22, Evinity Publishing Inc., ASIN B002CQU470. See also, **Kalpa Sutra**, by K.C. Lalwani, Motilal Banarsidass, 2nd edition, New Delhi, India, 1999. In addition, see **A Naked Man**, by Michael Tobias, Jain Publishing Company, Fremont, CA, 1994; see also Tobias' book, **Life Force: The World of Jainism,** 2nd Edition, Jain Publishing, Fremont, CA, 2000.

*100 "The Virgin among a Multitude of Animals," 1503, Pen and ink and watercolour on paper, 321 x 243 mm, Graphische Sammlung Albertina, Vienna. http://www.wga.hu/art/d/durer/2/11/2/03animal.jpg

*101 **The Complete Drawings of Albrecht Durer**, Vol. 2: 1500-1505, ed. Walter L. Strauss, Abaris Books, New York, p. 696. Jan Breughel was commissioned by the Emperor in 1604 to copy Durer's watercolor. See Kayo Hirakawa's **The Pictorialization of Dürer's Watercolors in Northern Europe in the Sixteenth and Seventeenth Centuries,** Peter Lang AG, International Academic Publishers, Bern, Switzerland, 2009, pp. 116-119.

*102 See **Megalopolis, the Urbanized Northeastern Seaboard of the United States**, M.I.T. Press, Cambridge, Massachusetts, 1966.

*103 See Joseph Meder's **Dürers Grüne Passion, Monographien Zur Deutschen Kunst**, Vierter Band, München, 1923.

*104 See **"Unto This Last:" Four Essays On The First Principles Of Political Economy,"** by John Ruskin, London: Smith, Elder And Co, 65, Cornhill, 1862. See "Cornhill Magazine – The John Murray Archive," National Library of Scotland, www.nis.uk/jma/topics/publishing/cornhill.html; See also, **The Complete Works of John Ruskin: Unto This Last**, **Munera Pulveris**, **Time**

and Tide with Other Writings on Political Economy 1860-1873, (17) by John Ruskin. E.T. Cook, and Alexander Wedderburn, Ballantyne Press, Edinburgh, Scotland, 1905. See also **Gandhi as Disciple and Mentor**, by Thomas Weber, Kindle Edition, January 10, 2005. Cambridge University Press, New York, 2005. See also, **Gandhi An Autobiography: The Story of My Experiments With Truth**, by Mohandas Karamchand (Mahatma) Gandhi, with Mahadev H. Desai, and Foreword by Sissela Bok, Beacon Press, Boston, MA, 1993 (first published in 1929).

*105 See **Sanctuary: Global Oases of Innocence**, by Michael Charles Tobias and Jane Gray Morrison, Foreword by Her Majesty Ashi Dorji Wangmo Wangchuck, Queen of Bhutan; A Dancing Star Foundation Book, 2008, www.sanctuary-thebook.org.

*106 See the essay, "Can Nature Make Us More Caring? Effects of Immersion in Nature on Intrinsic Aspirations and Generosity," by Netta Weinstein, Andrew K. Przybylski and Richard M. Ryan, Personality and Social Psychology Bulletin, Vol.35, No. 10, pp.1315-1329, 2009, at http://psp.sagepub.com/cgi/contnent/abstract/35/10/1315

Printed in Dunstable, United Kingdom

91222434R00100